Cities and Sexualities

From the hotspots of commercial sex through to the suburbia of twitching curtains, urban life and sexualities appear inseparable. Cities are the source of our most familiar images of sexual practice, and are the spaces where new understandings of sexuality take shape. In an era of global business and tourism, cities are also the hubs around which a global sex trade is organized and where virtual sex content is obsessively produced and consumed.

Detailing the relationships between sexed bodies, sexual subjectivities and forms of intimacy, *Cities and Sexualities* explores the role of the city in shaping our sexual lives. At the same time, it describes how the actions of urban governors, city planners, the police and judiciary combine to produce cities in which some sexual proclivities and tastes are normalized and others excluded. In so doing, it maps out the diverse sexual landscapes of the city – from spaces of courtship, coupling and cohabitation through to sites of adult entertainment, prostitution and pornography. Considering both the normative geographies of heterosexuality and monogamy, as well as urban geographies of radical/queer sex, this book provides a unique perspective on the relationship between sex and the city.

Cities and Sexualities offers a wide overview of the state-of-the-art in geographies and sociologies of sexuality, as well as an empirically grounded account of the forms of desire that animate the erotic city. It describes the diverse sexual landscapes that characterize both the contemporary Western city as well as cities in the global South. The book features a wide range of case studies as well as suggestions for further reading at the end of each chapter. It will appeal to undergraduate students studying Geography, Urban Studies, Gender Studies and Sociology.

Phil Hubbard is Professor of Urban Studies in the School of Social Policy, Sociology and Social Research, University of Kent, UK.

Routledge critical introductions to urbanism and the city

Edited by Malcolm Miles, University of Plymouth, UK
and John Rennie Short, University of Maryland, USA

The series is designed to allow undergraduate readers to make sense of, and find a critical way into, urbanism. It will:

- Cover a broad range of themes
- Introduce key ideas and sources
- Allow the author to articulate her/his own position
- Introduce complex arguments clearly and accessibly
- Bridge disciplines, and theory and practice
- Be affordable and well designed

The series covers social, political, economic, cultural and spatial concerns. It will appeal to students in architecture, cultural studies, geography, popular culture, sociology, urban studies and urban planning. It will be trans-disciplinary. Firmly situated in the present, it also introduces material from the cities of modernity and post-modernity.

Published:

Cities and Consumption – Mark Jayne
Cities and Cultures – Malcolm Miles
Cities and Nature – Lisa Benton-Short and John Rennie Short
Cities and Economies – Yeong-Hyun Kim and John Rennie Short
Cities and Cinema – Barbara Mennel
Cities and Gender – Helen Jarvis with Paula Kantor and Jonathan Cloke
Cities and Design – Paul L. Knox
Cities, Politics and Power – Simon Parker
Children, Youth and the City – Kathrin Hörshelmann and Lorraine van Blerk
Cities and Sexualities – Phil Hubbard

Forthcoming:

Cities and Climate Change – Harriet A. Bulkeley
Cities and Photography – Jane Tormey
Cities, Risk and Disaster – Christine Wamsler

Cities and Sexualities

Phil Hubbard

First published 2012
by Routledge
2 Park Square, Milton Park, Abingdon, Oxon OX14 4RN

Simultaneously published in the USA and Canada
by Routledge
711 Third Avenue, New York, NY 10017

Routledge is an imprint of the Taylor & Francis Group, an informa business

British Library Cataloguing in Publication Data
A catalogue record for this book is available from the British Library

Library of Congress Cataloging in Publication Data
Hubbard, Phil.
Cities and sexualities / Phil Hubbard. – 1st ed.
p. cm.
Includes bibliographical references and index.
1. Cities and towns–Social aspects. 2. Sex customs. 3. Social service and sex. I. Title.
HT119.H83 2011
307.76–dc22
2011011103

ISBN: 978–0–415–56645–2 (hbk)
ISBN: 978–0–415–56647–6 (pbk)
ISBN: 978–0–203–86149–3 (ebk)

Typeset in Times New Roman and Futura
by Florence Production Ltd, Stoodleigh, Devon

Printed and bound in Great Britain by
CPI Antony Rowe, Chippenham, Wiltshire

Contents

Case studies

Figures

Acknowledgements

This book has emerged from a number of collaborative projects and conversations over the last fifteen years. Notably, these have included projects with, among others, Maggie O'Neill, Jane Scoular and Jane Pitcher; writing collaborations with Baptiste Coulmont and Teela Sanders; and supervisory meetings with students working on the geographies of sexuality including Mary Whowell, Yu Chieh Hsieh and Jo Mitchinson. I wish to thank all of these for their support and inspiration over the years. The book also benefitted from the critical readings provided by Dennis Altman, Kath Browne, Natalie Oswin, Richard Phillips and Eleanor Wilkinson: though I have perhaps not addressed all their comments in the way they would have wished, their comments were stimulating and challenging, and have hopefully changed this text for the better.

Material in Chapter Seven has previously appeared in different form as a Globalization and World Cities Group bulletin on the Loughborough University Department of Geography website. I wish to thank my former colleagues at Loughborough for giving permission for sections of this bulletin to be reproduced here, as well as their support and advice over the course of ten years. Case Study 2.3 draws on ongoing research conducted as part of my Visiting Professorship at University Technology Sydney, and I wish to thank Spike Boydell, Jason Prior, Penny Crofts and others for involving me in the People, Place, Property and Planning project. Case Study 6.3 draws extensively on the research completed by Baptiste Coulmont which was incorporated in a jointly written paper (Coulmont and Hubbard 2010) and I must thank Baptiste for his patience with my frustratingly rudimentary French.

Unless otherwise stated, all photographs are by the author. However, I wish to thank John Harrison, Lauren Young and the Tate Collection for providing images for reproduction. All other images are licensed under Creative Commons agreements and authors have been credited as appropriate.

Preface

It begins with footsteps. Bare feet – a man's, then a woman's – leaving a bedroom and heading for a shower. The same shower? It's not clear. The scene splits. Shots of teeth being cleaned, hair gently tousled, make-up applied. Wardrobes open, towels drop, clothes are chosen. The wearers admire themselves in the mirror, she in a stylishly cut dress, he in open collar shirt, a black jacket. Two people getting ready; separate lives.

Cut to: a metro station in Paris. The woman and man enter the station, from either end, and stand on the platform, five metres or so apart. Two strangers, standing alone. The man glances at the women, looks her up and down. She notices, smiles back. The train arrives.

On the train, they are sitting in the same carriage, alone and silent. The man ignores the woman, but glances in the window where he catches a glimpse of her legs, crossing and uncrossing. He starts to imagine what those legs look like when she is undressed. He turns away. She looks across, noticing him again, and imagines the curve of his back.

A montage of shots follow: the man and woman talking in a square by a fountain; caressing by the River Seine late at night; in a nightclub, drinking; in a cinema; in a restaurant: a glamorous couple, in love.

The woman's phone rings, breaking this sequence and returning us to the metro. Was this her fantasy? Or his? We can't tell. Yet the phonecall attracts the man's attention back to her, and he starts to imagine her with another man, plain-looking. He laughs to think that someone so beautiful is with someone like that, and not with someone like him.

Another metro stop passes. The woman smiles at him, and we see more images. His cigar burning a mark on a carpet; a row on the street; getting into a nice car; getting into a cheaper car; rows in traffic; driving to their flat; a ring on a finger.

She smiles at him, he smiles again. A final montage: the couple on a bed, touching, caressing; her dress falling from her shoulders; her hand stroking

> his neck; passionate kisses; feet entwining; the couple naked, fucking, him on top, then her.
>
> The metro draws to its destination, shaking them from their reverie, they get up, move to the door, he behind her. We see their reflection in the metro door, where they appear as a couple. But when the doors open, they depart in different directions, two separate lives.

Florian Sela's short film – *L'amour dure Trois Minutes* – ('Love Lasts Three Minutes') is a play on Frederic Beigbeder's (1997) novel *Amour dure Trois Ans* ('Love Lasts Three Years'), and charts an encounter between two protagonists who, drawn by mutual attraction, fantasize about a (sex) life together. No words are exchanged, no bodies touch, yet in the movement of an eye, a slight smile, a shift in posture, we can sense that there was the possibility that this encounter could have been more than it was, just one of the thousands of random encounters that animate everyday urban life.

In drawing attention to the possibilities of the city as a space in which to find and experience sex, romance and love, Sela's film effectively underscores several of the themes that this book explores. One is that cities – and especially big cities – are sites where disconnected people, perhaps from different cultural and geographical backgrounds, are drawn into sexual relationships bound by the rules of attraction. These relationship can be real or imagined, and are of varied duration – from a one-night stand to a lifetime of companionship – but are united by the fact that they involve people sharing forms of personal and bodily intimacy. This means that their relationship moves from the realm of normal sociality and friendship into a 'socio-erotic' domain. Of course, this does not necessarily mean that their relationship is any more meaningful than any other of their relationships or acquaintances, and it might not be characterized by any element of emotional reciprocity or empathy, but it certainly means that their relations are experienced as being of a sexual nature. As such, the city can be seen as a site whose inherent heterogeneity and difference is bound together through the promise, pursuit and practices of sex itself.

Sela's film is telling to the extent that it shows how strangers can come together, drawn by sexual attraction, within a city of millions. It shows that even within the confines of our banal urban lives – a three-minute journey – the city provides a stage for performances that are sexual or sexualized. The public spaces and streets of the city, in particular, are spaces where people dress, walk and talk in ways that they hope will attract the sexual attention of others: attracting welcoming glances while repelling the gaze of the unwanted and undesired is a key urban skill, involving forms of body- and face-work that render bodies coherent and desirable to selective audiences. Other pseudo-public spaces, such as retail, office and leisure spaces, can also be important as spaces where bodies are, to lesser or greater extents, being appraised in sexual terms, and where people modify their dress and appearance according to how they might like to be perceived. This is not to say that bodies are not sexualized or sexy in the rural – far from it – but that the sheer density and busy-ness of city life means that the city offers multiple possibilities for sexual encounter, and provides a theatre for sexual display.

Significantly, Sela's film also perpetuates the idea that certain cities are inherently sexy. It seems reasonable to assume that two passengers on a metro in Paris are no more likely to be attracted to one another than if they are sitting on a metro in Tokyo, Moscow or Newcastle. Yet for all this, Paris has a reputation as a city for lovers, and is celebrated as a honeymoon destination (the Eiffel Tower is known as one of the world's most popular spots for people to propose to one another). Just exactly what it is about Paris that makes it romantic is difficult to discern, though clearly media images and advertising exploit this reputation. But irrespective of such representations, it is unquestionable that Paris's boulevards, cafés, nightspots, parks and restaurants feel – or have been made to feel – sexy and chic. In short, the city has an ambience that lends itself to sexual encounter, and remains a major focus for forms of 'sexy tourism' and sexual commerce.

Putting some of these themes together, we can begin to understand why there is a need for a book considering the relations between sexuality and the city. Put simply, cities have long been recognized as spaces of sexual encounter, as sites where bodies come together, mix and mingle. They are known as sites of sexual experimentation, radicalism and freedom, as places where individuals can pursue or purchase a rich diversity of sexual pleasures. Historically, the city has been regarded as a space of social and sexual liberation because it is understood to offer anonymity and an escape from the more claustrophobic kinship and community relations of smaller towns and villages. For such reasons – and not just the lure of work – young single people tend to congregate in cities, making urban areas the most vibrant of all settings in which people search for sex partners. This is understood to be especially important for individuals identifying as lesbian, gay or bisexual, who historically have gravitated towards larger cities:

> Cities offered a larger selection of partners than smaller towns and villages. Crowds provided anonymity and, where homosexual acts remained illegal, a measure of safety. Migrants could break out of the strictures imposed elsewhere, locating new 'sub-cultures' to satisfy reprobate desires . . . Libido, hope for friendship and romance, and a need for money, drove

> them to search out casual, situational or long-term partners or patrons. Cities have provided venues where men who have sex with men (and women who have sex with women) can meet: pubs and clubs, cafes and cabarets. In times of clandestine homosexuality, public baths and toilets, parks and back streets were especially hospitable to trysts.
>
> (Aldrich 2004, 1725)

Often contrasted with a rurality that is deemed sexually conservative and even backward, the city is hence widely regarded as a site of sexual liberation, with key cities (such as Paris, but also San Francisco, London, Amsterdam, Sydney and New York) playing host to visible lesbian and gay communities characterized by alternative, extended and voluntary kinship patterns (Adler and Brenner 1992; Forest 1995; Knopp 1998; Podmore 2006; Doderer 2011). This association between lesbian and gay sexuality and the metropolis is effectively captured in the title of Weston's (1995) article about the gay geographical imagination: 'Get thee to a big city'. Cities are also sites where both the meanings and practices of sex itself have been transformed and, in recent times, where particular forms of sexual consumption and leisure have been normalized and encouraged. Sexual commerce abounds in most (if not all) cities, whether visibly in the form of lap dance clubs, peep shows, sex shops and areas of street sex working, or in more clandestine and privatized forms off-street. Whatever one's sexual predilections, the city seemingly provides places where one's desires can be fulfilled (albeit often at a price).

Yet this representation of the sexually liberal city does not tell the full story, as while the metropolis has been a notable location of sexual experimentation, it has also been the site where sexuality is most intensely scrutinized, policed and disciplined. It is a location where sexual orders have been worked and reworked, and where ideas of the 'normal' and 'perverse' have been both instituted and contested. This is because cities are not just comings-together of people in the interests of social and economic reproduction; they are also sites of governance from which power is exercised through various apparatuses of the state. Cities are indeed host to the key institutions that have a vested interest in regulating sex as part of a project of maintaining social order: the police, local government, departments of planning and housing, the courts, hospitals, probation services, social services and so on. These institutions are rarely discrete or isolated, extending their reach from metropolitan centres out to the provinces and the countryside through *geographies* – as opposed to simply *geometries* – of power (Howell 2009).

All of this is to suggest that the relationship between cities and sexualities is ambivalent: the city both enables and constrains sexuality, intensifying desire and repressing it in different ways. However, to state this is not enough, for it fails to grasp what is truly at stake in the relationship between sex and the city: as Mort and Nead (1999, 7) argue, to maintain that space is constitutive of sexuality is at once to 'say everything and nothing at all' given this is a broad, theoretical abstraction. Mort and Nead instead contend there is a need to examine how the city distributes and regulates sexual subjects and populations, and to explore how those subjects and populations negotiate and live out their sexual lives *within* urban spaces. The city is therefore far from being a neutral backdrop against which sexual relations are played out: it is an active agent in the making of sexualities, promoting some and repressing others. To map the urban geographies of sex is to expose the ways in which sexuality is subject to discipline and the exercise of knowledge and power. To put this in simple terms: each time sex takes place, and occupies space, it territorializes a particular understanding of sexuality (see Perreau 2008). To paraphrase Michel Foucault, a history of sexualities is therefore a history of spaces.

As Houlbrook (2006) insists, the city and sexuality are thus conceptually and culturally inseparable. In the public imagination, the association between sex and the city is axiomatic: for example, television crews cruise the city's streets seeking to pair people up (*Streetmate*), expose the diverse sex scenes of different cities (*Sexcetera*) and celebrate the sexual exploits of the city's inhabitants (most famously, perhaps, in *Sex and the City*, whose representation of sexually voracious women pursuing pleasure in New York cemented its reputation as an enlightened city of sexual opportunity). Yet in academic circles, there are still many who explore the dynamics and socialities of human sexuality without considering the importance of *space* and *place* in the making of our sexual lives (not to mention those who even seek to strip sexuality of its social and cultural context, considering it as a solely biological imperative). This book shows that exploring the sex lives of the city without noting the role of the city in shaping those sexualities is to miss a crucial dimension of human sexuality: the city is not simply the context for sex but plays an active role in the shaping our desires. Conversely, this book also argues that we cannot hope to understand the city without considering the importance of sexuality. Any urban theory that does not acknowledge the importance of sexuality ignores a vital dimension of social life – and one of the key factors that shapes our experience of the city.

Detailing the relationships between bodies, places and desires, this book accordingly explores the role of the city in shaping our sexual lives. At the same time, it describes how the actions of urban designers, planners and governors produce particular types of city in which some sexual predilections and tastes are catered for but others excluded. Highlighting the inescapable relationship between sex and the city, and emphasizing its gendered dimensions, it thus presents a series of linked case studies that explore how the city celebrates or represses particular sexual desires. Unlike many works in the field, it does not seek to privilege a specific set of sexual relations or identity positions but takes a broad overview of human sexualities. This will entail a consideration of the urban lives of those who identify as heterosexual, lesbian or gay as well as those populations who less readily identify with these discrete sexual identity positions (e.g. those who identify as bisexual, or resist labelling altogether). Importantly, the book will also move beyond questions of identity to consider questions of *practice*. Here, analysis of sex work, the consumption of pornography, the pursuit of anonymous sex and non-monogamous lifestyles will suggest that some sexualities are socially and spatially marginalized, while other sexualized practices (and bodies) are made to appear normal through their repeated and ubiquitous performance in urban space.

While there are now a number of important collections that explore the relationship between sexuality and space (for example, Bell and Valentine 1995a; Browne *et al.* 2007; Johnston and Longhurst 2010), this book aims to make a significant intervention in the cross-disciplinary field of sexuality studies. It seeks to do this by moving beyond binaries of straight/gay identity, offering a wider overview of the ways that the city shapes our sexual and intimate lives, encouraging us all to perform sexualities that are, in some sense, 'normal' and comprehensible within the socially dominant models that suggest sex is something that is most appropriate, and rewarding, within the context of a consensual, and preferably loving, long-term relationship (Hubbard 2000). To make these arguments, it draws on a diverse range of empirical material, using case studies to explore particular spaces of sex within the contemporary Western city as well as some cities in the global South. These case studies are, by their very nature, limited in scope and depth, but are designed to be more than simply illustrative: they are intended to inspire readers to further explore these – and other – stories of sex and the city.

1 Introducing cities and sexualities

Learning objectives

- **To understand that sexuality is simultaneously biological, psychological and social.**
- **To appreciate why place matters in an understanding of sexualities.**
- **To gain an understanding of some of the key processes that serve to sexualize the city.**

Traditionally, sexuality has been of relatively little interest to urban researchers, who have appeared remarkably reluctant to explore the way that people's sexuality shapes, or is shaped by, their urban experiences. There are a number of possible explanations for this, aside from the general prudishness that is often evident about sex. One possible explanation is that sexuality, if conceived in its narrowest biological sense, can be seen to concern our sexual behaviour and the physiological and psychological basis of our sexuality. Viewed in this way, sex might be understood as a solely biological imperative, worthy of investigation by the clinician, the medical professional or the sex therapist, but something that seems to be little influenced by a person's surroundings. Whether one is born, or lives, in a sprawling metropolis or tiny rural hamlet seems to have little bearing on the materiality of the body, or the sexual desires that we possess, given we are born into bodies that determine our subsequent sexual development.

However, if viewed from an alternative, sociological perspective, sex can be regarded as the product of social forces that need to be explained, with people's sexuality shaped by their gender, age, class and ethnicity, as well as the cultural influences to which they are exposed in their everyday lives. A sociological perspective hence views sex as not something dictated by our physical needs and urges, but embedded in multiple institutions, networks and organizations that shape our desire. This does not mean that social science approaches ignore the biology or physicality of the body. Far from it. Sex itself is always embodied and visceral, involving fleshy, desiring bodies, touches, looks, tastes, smells, bodily fluids, sperm, saliva, sweat. But sociologists argue that the embodied experience of sex can never transcend the social, with sex always being informed by the images of eroticism that circulate in the media, the conversations we have about sex and the guidance we are given about what sex is supposed to be. When we have sex, or claim a sexual identity, we are thus positioned within the social.

While not discounting ideas about the science of sex, this book is hence grounded in social science literatures that explore the *social construction* of sex. Such literatures suggest that while sex is always a matter of biology (i.e. embodied acts and physical processes), the fact that specific bodily actions and performances are understood as 'sexual' or 'erotic' means they take on a *meaning* that ripples out to encompass all dimensions of our identities and practices. To take an example: virginity might be understood biologically as having not had penetrative sex with another person, but socially it is surrounded by a complex range of assumptions and understandings of purity, cleanliness and innocence. Moreover, there are seen to be 'right' and 'wrong' ages at which to start having sex, and 'good' and 'bad' ways to lose one's virginity. These social myths and meanings have important consequences for what it means to be a virgin, and what it *feels* like to be a virgin. Moreover, they also have an important influence on decisions to 'lose' one's virginity, or perhaps to perform a celibate identity that celebrates the decision to remain a virgin in the face of social pressures to the contrary (Abbott 2000).

This type of example suggests that not all sexual acts or identities are regarded as equivalent in contemporary society. As Gayle Rubin (1984) argued in her essay 'Thinking Sex', some sexualities are socially privileged, others marginalized. As such, it is possible to speak of 'good sex' – that which the state, media and law suggests is normal, natural and healthy – as well as 'bad sex' – that which is depicted as 'utterly repulsive and devoid of all emotional nuance' (Rubin 1984, 117). Writing in the context of the mid 1980s, Rubin (1984, 117) argued that the latter encompassed the 'most despised sexual castes . . . transsexuals, transvestites, fetishists, sadomasochists, sex workers such as prostitutes and porn models, and the lowliest of all, those whose eroticism transgresses generational boundaries'.

In situating particular acts and identities as immoral, and thus on the 'margins' of acceptability, the moral 'centre' is defined. The boundaries between moral/immoral and good/bad sex are never clear cut, however, with changing understandings of sex being circulated, and contested, via social representations of different sexual practices and lifestyles. Some sexualities have shifted from being 'bad' to 'good': for example, while Rubin spoke of homosexuality's marginal and even criminalized status in the 1980s, this has been transformed by the efforts of homophile and, later, queer activist groups since that time, with many nations now recognizing same-sex civil partnerships and offering lesbian and gay identified individuals protection from homophobic discrimination and abuse (McGhee 2004). By the same token, however, sexualities can move from the centre to the margins of society: for example, in classical times, it appears that Athenian society revolved around male-dominated and homosocial notions of bonding that encouraged older men to take younger boys – of between twelve and eighteen – as lovers (Halperin 2002; Clark 2008). Today, as Rubin notes, such behaviour would be widely condemned as paedophilia.

Rubin (1984, 116) hence argued that 'all erotic behaviour is considered bad unless a specific reason to exempt it has been established, with the most acceptable excuses [being] marriage, reproduction, and love'. This argument has subsequently been developed by researchers exploring how a particular, coupled, form of heterosexuality is made to appear natural and normal, something captured in the concept of *heteronormativity*:

> By heteronormativity we mean the institutions, structures of understanding, and practical orientations that make heterosexuality seem not only coherent – that is, organized as a sexuality – but also privileged . . . It consists less of norms that could be summarized as a body of doctrine than of a sense of rightness produced in contradictory manifestations – often unconscious, immanent to practice or to institutions. Contexts that have little visible relation to sex practice, such as life narrative and generational identity, can be heteronormative in this sense, while in other contexts forms of sex between men and women might *not* be heteronormative. Heteronormativity is thus a concept distinct from heterosexuality.
>
> (Berlant and Warner 1998, 178)

This concept of heteronormativity has proved important given it develops Rubin's idea of a hierarchy of sexualities and explores the normalization not of heterosexuality per se, but a form of heterosexuality based on coupling, reproduction, consensual sex and love. This is also the form of sexuality privileged by the state, with most nations granting certain rights of citizenship to coupled, reproductive individuals which are denied to 'bad' sexual subjects (Richardson 2000).

Exploring shifts in social understandings of what is sexually 'normal' is important in any examination of sex and the city, for it underlines that understandings of what sexuality is – and which sexualities are 'normal' – can vary across both time and space. Despite the fact that we live a global world, where there is some degree of cultural homogenization, it is obvious that there are different understandings of 'appropriate' sexual comportment and manners between East and West, and between the global South and the global North, with significant variations apparent within these broadly defined areas (see Hastings and Magowan 2010). This suggests that although it is possible to make generalizations about the sexual life of cities, there are certain dangers in imagining that all cities promote the same sort of sexualities (see Brown *et al.* 2010 on urban sexualities beyond the West). Bearing this in mind, this chapter will begin to trace the connections between sex and the city by exploring how nineteenth-century European urbanization triggered new understandings of 'good' and 'bad' sexuality. The first section hence explores how the emergence of large, modern cities (such as London, Berlin and Paris) prompted anxieties about the sex lives of their citizens. A key idea in the second section of this chapter is that these anxieties fuelled attempts to order the city via acts of planning, environmental modification and health reform which were ultimately about disciplining the city's diverse sexualities. Acknowledging that such acts have tended to produce heteronormative cities, the final section of this chapter stresses that the city nonetheless remains a site where sexual norms can be questioned or exceeded, offering diverse spaces for the performance of alternative, residual or 'queer' sexualities. Urban space is hence shown to be highly significant in shaping the sexual life of its citizens, distributing bodies and desires to produce cities where particular forms of sexual conduct dominate: as Mitchell (2000, 35) notes, 'like any social relationship, sexuality is inherently spatial – it depends on particular spaces for its construction and in turn produces and reproduces the spaces in which sexuality can be, and was, forged'.

Diversity and danger: urbanization and sexual anxiety

While the first cities emerged thousands of years ago, it was only in the nineteenth century that the city began to be taken seriously as a distinctive and important academic object of study. One of the main reasons for this was that, until that time, the overall share of the global population living in cities was small in both absolute and relative terms. The rapid urbanization of the nineteenth century changed this, evident first and foremost in the economically dominant states of Europe and then in the cities of the so-called 'New World', notably the US. What was particularly significant about this process of urbanization was that it produced cities that contemporary commentators struggled to describe using existing language: their size, appearance and apparent complexity rendered them a new *species* that demanded to be classified, catalogued and ultimately, diagnosed.

In its nascent form, urban studies was concerned with describing the distinctive social, economic and political life of these cities, noting they were more crowded, diverse and individualized than rural settlements. The idea that the city represented the antithesis of traditional ruralism became particularly associated with Ferdinand Tonnies' (1887) distinction between *gemeinschaft* communities – characterized by people working together for the common good, united by ties of family (kinship), language and folklore – and *gesellschaft* societies, characterized by rampant individualism and a concomitant lack of community cohesion. Though Tonnies couched the distinction between *gemeinschaft* and *gesellschaft* in terms of a pre-industrial/industrial divide rather than an urban/rural one, his description of *gesellschaft* societies was deemed appropriate for industrial cities where the extended family unit was supplanted by 'nuclear' households in which individuals were concerned with their own problems, and seldom those of others, remaining indifferent even to those in their immediate neighbourhood.

Though caricatured, the idea that urban settlements were less cohesive than their rural counterparts was a persuasive one, and resonated with discourses that figured the modern city as cold, calculating and anonymous. Friedrich Engels' (1844) work is of particular note in this respect given it documented the *inhuman* living conditions experienced by workers in cities that increasingly served the interests of industrial production and the property-owning classes. In a more general sense, Georg Simmel (1858–1918) described the impacts of urbanism on social psychology, suggesting that the city demanded human adaptation to cope with its size and complexity. In his essay 'The metropolis and mental life', Simmel (1903) argued that the unique trait of the modern city was the *intensification* of nervous stimuli with which the city dweller must cope. Describing the contrast between the rural, where the rhythm of life and sensory imagery was slow, and the city, with its 'swift and continuous shift of external and internal stimuli', Simmel detailed how individuals psychologically adapted to urban life. Most famously, he spoke of the development of a blasé attitude – the attitude of indifference which urban dwellers adapt as they go about their day-to-day business (something that remains evident in the etiquette of urban life, where adopting modes of 'civil inattention' enables the pedestrian to negotiate encounters with the innumerable strangers passed in the street) (Smith and Davidson 2008).

The idea that the urban experience is essentially 'managed' through a transformation of individual consciousness that involves a filtering out of the detail and minutiae of city existence remains an important foundation for urban theory. So too does the idea that city life debases human relations, and renders contact between urban dwellers essentially superficial, self-centred and shallow, based on surface appearance. For Simmel, the impersonality and depthlessness of urban life was related to the fact that the industrial city served the calculative imperatives of *money*. Simmel essentially suggested this encouraged relations based purely on exchange value and productivity (and thus dissolved bonds constructed on the basis of blood, kinship or loyalty). This, he argued, encouraged a purely logical way of thinking which valued punctuality, calculability and exactness. The corollary was a city that moved to the rhythms of industrial capitalism, and was marked by a ceaseless transformation (Berman 1983). This was to have important consequences for the sex life of modern cities, as Brown and Browne summarize:

> These new forms of urban life and the anonymity and freedom afforded by large, concentrated populations enabled unorthodox sexual practices and the development of new subcultures based around minority sexualities . . . [with] sexual adventure to be found in the circulation of the crowd, the comingling of different classes in public space, and the spectacle of the electrified city at night.
>
> (Brown and Browne 2009, 697)

Despite an evident reticence to situate sexuality within the realm of the social, the pioneers of urban sociology began to note that the great metropolitan centres were characterized by distinctive sexualities (Heap 2003). This led to the development of numerous theories linking sexual 'perversion' to the social turbulence and disorganization of the modern city (see Case Study 1.1).

The attributes deemed characteristic of modern cities – anonymity, voyeurism, exhibitionism, consumption, tactility, motion and restlessness – hence played a role in facilitating a more diverse range of sexual behaviours than those evident in traditional rural societies. Freed from the constraints of rural kinship networks, the urban dweller could explore new sexual avenues and negotiate new sexual identities (something clearly evidenced in the example of Berlin, where both women and men found multiple opportunities for sexual encounter outside the traditional confines of marriage). For Iris Marion Young (1990, 224), city life began to *eroticize* difference 'in the wider sense of an attraction to the other, the pleasure and excitement of being drawn out of one's secure routine to encounter the novel, strange and surprising'. The coming-together of individuals from different backgrounds and origins therefore not only provided the possibility of forging new sexual identities

CASE STUDY 1.1

Sex in the metropolis: Weimar Berlin

The third largest city in the world after London and New York, Berlin transformed dramatically in the boom years of the 'Golden Twenties' between the inflation crisis of November 1923 and the Wall Street Crash of October 1929. One symptom of this was the emergence of an extensive night-life district centred on a diverse range of theatres, opera and, most famously, cabaret clubs. Another was creativity and experimentation in art, literature, design and architecture: the outward appearance of the city betrayed this, being characterized by a new 'objective' style of architecture – pioneered by Gropius, Taut, Mendelsohn and Mies van der Rohe – which made a virtue of structural integrity, functional, clean appearance and lack of ornament. Berlin thus became known as a truly *modern* metropolis, thoroughly of the moment: in the words of journalist and social commentator Siegfried Kracauer (1932, cited in Frisby 2001, 64), 'it appears as if the city had control of the magic means of eradicating memories. It is present day and makes a point of honour of being absolutely present day.'

Summarizing the socio-spatial transformation of 1920s Berlin, Ward (2001) comments on its evident 'surface culture', suggesting that the new objectivity hid nothing. All was on display, so to speak. From its department stores to the streets themselves, the city offered an excess of commodities that were fashioned, packaged and displayed in an aesthetic manner to increase their 'external appeal', and Ward suggests that this extended to the body itself. In its commodified form, the body took on the attributes of the city, with bodies culturally – and economically – valued for their efficient, modern, stylish appearance. This was most evident in the emergence of new fashions for men and women, particularly those associated with the *Neue Frau* ('New Woman') whose short bobbed hair, penchant for smoking and relaxed, almost masculine clothes, emphasized the new-found freedoms of the modern city and the sexual liberation this implied. Indelibly associated with the Berlin cabaret scene – captured in Christopher Isherwood's (1939) *Goodbye to Berlin*, and the film it inspired (Bob Fosse's *Cabaret*, 1972) – the New Woman was located imaginatively in the decadent all-night clubs that satirized dominant political mores and often flouted normal conventions around nudity and dress. In such settings, the *Neue Frau* was presented as an object to be visually and sexually consumed by men, despite her education and evident mobility.

The idea that the modern city accentuated the visible and the visual through an exaggerated and intense emphasis on surface form suggests Berlin was a stage where new sexualities were not just performed, but *produced*. Writing in 1903 on the processes of commodification and spectacularization associated with the rise of modern urban culture, Berlin-based sociologist Georg Simmel argued that human emotion was being reduced to a sexual and economic exchange, with the traditional ties of kinship being subsumed by a more individualized culture that allowed for sexualities that transcended traditional gender, racial and class boundaries. One obvious symptom in Weimar Berlin was the development of well-known 'sex zones' where as many as 30,000 prostitutes worked after the decriminalization of sex work in 1927. The most infamous 'red light area' was that around Alexanderplatz, which contained upwards of 300 brothels, but throughout the city streetwalkers signified they were for sale through provocative modes of dressing. Overlapping these sites of prostitution, Berlin was also host to some of the first openly gay and lesbian bars (with the foundation of gay rights organization Berliner Freundesbund in 1919) (Evans 2003; B. Smith 2010).

Internationally, Berlin hence became known for its sexual experimentation and liberalism. The mood was well captured by George Grosz (1893–1959) and Otto Dix (1891–1969), artists whose caricatures and paintings provided some of the most vitriolic social criticism of their time. They depicted a city populated by prostitutes, bloated businessmen and disfigured war veterans (see Figure 1.1). Tellingly, many of their paintings were of the act of *lustmord* – the sexual murder of women. Though Berlin was not especially characterized by sexual crimes of this type (in fact more occurred in the rural), the representation of murdered women functioned as 'an aesthetic strategy for managing certain kinds of sexual, social and political anxieties' (Tatar 1995, 76). Retrospectively, it has been argued that such depictions of Berlin as decadent and depraved were especially associated with male fears of a city reigned over by the free-roaming libido of women, and a sexual laxity celebrated in the nightclub scene (Rowe 2003). Equally, emerging sexological discourse suggested that while men were naturally polygamous, the city was causing more widespread moral corruption of women: 'The problem lay with the city itself, whose anonymity, artificiality and rampant commercialism over-stimulated the libido and distorted the balance of nature within and between the sexes' (Forel and Fetscher 1931, 34).

Ultimately, it has been suggested that such exaggerated reactions to the 'threat' of untrammelled female sexuality figured in the censorship and repression of the cabaret scene under the Nazis in the 1930s. However, the flowering of sexual diversity in the Berlin of the 1920s underlines that the city is no mere backdrop against which our sexual lives are played out, but can play an active role in the production of sexuality.

Further reading: Ward (2001); B. Smith (2010)

Figure 1.1 ***Suicide*****, George Grosz, 1916 (courtesy of the Tate Collection).**

in the midst of an anonymous crowd, it also provided the opportunity for challenging traditional ideas of sexual conduct. However, few urban scholars paused to dwell on this at the time, beyond noting forms of 'sexual immorality' that dwelt in 'bohemia, the half-world and the red-light district' (Park 1915, 612): often it was assumed that immorality flourished in the city's *transitional zones*, populated by immigrants, working class itinerants and street people (see Chapter Two).

The connection between 'sexual deviance' and the city remained a significant focus of urban sociology and ethnography throughout the twentieth century (Kneeland and Davis 1917; Cressey 1932; Reckless 1933; Symanski 1981). However, it was not until the 1980s that wider issues of sexuality began to feature prominently in urban studies, particularly in literatures exploring the emergence of lesbian and gay neighbourhoods (Levine 1979; Weightman 1980; Castells 1983), as well as in various ethnographies of sexual meeting grounds (Humphreys 1970; McKinstry 1974; Warren 1974). Subsequently, research exploring the relationship between sex and the city has proliferated, encompassing studies not just of the urban geographies of lesbian and gay identified individuals, but the sex lives of a more diverse range of urban dwellers, whether they identify as straight, lesbian, gay, bisexual, monogamous or non-monogamous (for overviews, see Binnie and Valentine 1999; Knopp 2007; Hubbard 2008; Perreau 2008; Johnston and Longhurst 2010). This has often blurred into discussions about the gendering of urban space, with emerging work on transgendered spaces posing important questions about the assumed connections between sex and gender identities (Doan 2004; Browne and Lim 2010).

One recurrent theme in the literature is that the materialities of the city encourage sexual encounters by intensifying desire (Brown 2008b; Johnston 2010). For example, the lighting of the city masks and reveals different sights in ways that can evoke anticipation, anxiety or longing (see Chapter Five), while ever-present images of sexualized bodies on billboards, signage and advertising hoardings effectively remind viewers that the city is a sexual marketplace where bodies are constantly on display and all is for sale (Rosewarne 2007). Such examples stress that *vision* is crucial in the sex life of the city and the making of urban sexual subjectivities (Pile 1996). Yet the 'erotics of looking' (Bell and Binnie 1998) must be considered alongside other sensations and experiences of the urban – for example, the ways cities sound, smell, taste and feel – and the ways these haptic and sensory geographies are connected to the movements of bodies, the rhythms of streetlife, the appearance of buildings, urban microclimates, the design of public spaces and so on. Combined, these can effectively *sexualize* space:

> Space and place work together in the formation of sexual space, inspiring and circumscribing the range of possible erotic forms and practices within a given setting. The atmospheric qualities of a given locale are thus both hard and soft, immediate and (potentially) diffuse: location . . . architecture, décor, history and site-generated official and popular discourses merge into a singular entity (though there may be multiple interpretations of it). The immediate properties of a given space's atmosphere suggest to participants the state of mind to adopt, the kinds of sociality to expect and the forms of appropriate conduct. They also facilitate or discourage types of conduct and encounter.
>
> (Green *et al.* 2008, 5–6)

Iris Marion Young (1990, 224) made a similar point when she argued that 'the city's eroticism . . . derives from the aesthetics of its material being, the bright and coloured lights, the grandeur of the buildings, the juxtaposition of architectures of different times, styles and purpose . . . its social and spatial inexhaustibility'. The idea that cities are eroticized 'via the constant titillation of the senses by sex' (Mumford 1961, 24), has given rise to one of the enduring dimensions of the imagined urban/rural binary: the notion that the countryside is staid, conservative and moral in its sexualities, whereas the city is sexually experimental, liberal and promiscuous (but see Phillips *et al.* 2000). Depending on one's perspective, the city can be a sexual utopia or a site of unrelenting sin and immorality, its importance as a place of bodily contact and pleasure-seeking having being used to prop up anti-urban myths of the city as Sodom, as well as bolstering a pro-urban sentiment connecting sexual freedom to the emancipatory politics of the city. This latter emphasis is one perpetuated in much academic writing on urban sexuality, particularly accounts which highlight the pivotal role of the city in the making of lesbian and gay communities (d'Emilio 1998; Aldrich 2004; Chisholm 2005; Abraham 2009). But while such literatures note the city's apparent capacity to accommodate sexual difference, they also note the limits placed on that freedom, and the curtailment of particular sexual pleasures. Accordingly, a key theme in studies of sexuality and space is that the city is a key site in the control and *disciplining* of sexuality.

Disciplining the metropolis: regulating sexuality, spatially

Urban histories of Berlin, Paris, London, New York and the other metropolitan centres of the nineteenth century suggest cultural anxieties about sex intensified in response to the rapid social and economic changes associated with urban growth. Put simply, linked processes of urbanization (increasing numbers living in cities), modernization (the increasing importance of new scientific, technological and cultural innovations) and secularization (the associated decline of religious influence) created large urban centres where individuals appeared more able to pursue more diverse sexual pleasures. Coupled with this, innovative architectural forms, spectacular spaces of leisure and new topographies of urban space provided an ambience more conducive to sexual encounter, providing multiple spaces of seduction and raising the spectre of inappropriate sexual mixing (Howell 2001; Cook 2003). Rapid urbanization in the nineteenth and early twentieth century therefore seemed to be associated with shifts in sexual sensibilities (a trend that seems to hold true in the context of the hyper-urbanization experienced in China and East Asia in the twenty-first century, where new spaces of sexual identification appear to be opening up – see Farrar 2002; Zheng 2009).

Periods of rapid urbanization hence appear to be times when sexual life shifts in significant ways. Often, these changes are disorientating, requiring the production of new sexual knowledges. Indeed, the identification of 'sexuality' as a specific, erotic aspect of human life was arguably something that was triggered by the rise of modern, large cities. This is not to deny that sex took diverse forms in smaller, pre-modern cities, with different taboos concerning particular types of sex evident long before modernization took hold. Rather, following the work of Michel Foucault, it is to insist that sex began to develop a series of specific meanings beyond its sheer physicality as new, modern ideas emerged about the relationship between pleasure and carnality. These ideas were traced in Foucault's unfinished three-volume history of sexuality, which described how, in pre-industrial European societies, sexual practices were primarily subjected to moral and religious scrutiny and categorized in relation to sin, with non-reproductive sex represented as immoral. In contrast, Foucault described how in the modern, industrial era, varied 'technologies of sex' began to explore sexuality both within and beyond marriage, subjecting a more diverse range of bodily acts and practices to scrutiny as part of 'scientific' consideration of people's sexual lives that took shape at the interface of psychology, psychiatry, psychoanalysis and psychotherapy. Sexology – the 'science of sex' – accordingly emerged in the latter years of the nineteenth century as a set of diagnostic practices and knowledges that sought to understand sex as motivated by, and related to, pleasure. This tended to conceptualize sexual behaviour as the outcome of physiognomic drives, with scientific research exploring the biological basis of different sexualities, whether normal or 'deviant'.

Foucault hence examined the histories of sexuality as a concept, noting that, unlike sex itself, sexuality is a cultural production that *appropriates* the human body and its physiological capacities, connecting it to particular ideas about sex and gender:

> Sex must not be thought of as a kind of natural given which power tries to hold in check, or as an obscure domain which knowledge tries gradually to uncover. It is the name that can be given to a historical construct [*dispositif*]: not a furtive reality that is difficult to grasp, but a great surface network in which the stimulation of bodies, the intensification of pleasures, the incitement to discourse, the formation of special knowledges, the strengthening of controls and resistances, are linked to one another, in accordance with a few major strategies of knowledge and power.
>
> (Foucault 1978, 105–6)

Foucault's consideration of the emergence of modern discourses of sex has subsequently been massively influential, and spawned many 'Foucauldian' analyses of how sexuality is connected to the exercise of knowledge and power. Such analyses have often explored the way that the state – through varied techniques of governance – has sought to impart particular ideas of appropriate sexual conduct. For example, there have been multiple studies of how Victorian science constructed certain 'truths' about safe sex, health and disease, allying these to particular ideas of psychological happiness and development (Walkowitz 1982; Weeks 1985; Halperin 1986; Birken 1988; Hall 1992; Mort 2000). Such analyses have considered how sex was 'policed' in the nineteenth century through a combination of medical and moral ideas which suggested that sexual behaviour needed to be subject to state surveillance as well as an 'auto-surveillance' exercised by individuals themselves, who were encouraged to regulate their own behaviour in the interests of the social good. Sex was no longer an issue that concerned just the well-being of the individual, but society as a whole, whose 'future and fortune were tied . . . to the manner in which each individual made use of his sex' (Foucault 1978, 25).

The increasing importance placed on sexuality by the state in the nineteenth century made possible an unprecedented power and control over both individuals and populations through the exercise of *biopower* – the 'numerous and diverse techniques for achieving the subjugations of bodies and the control of populations' (Foucault 1978, 140). Foucault (1990, 23) consequently highlighted the importance of sexuality in the reproduction of the state, noting that:

> sexuality is not the most intractable element in power relations, but rather one of those endowed with the greatest instrumentality: useful for the greatest number of maneuvers and capable of serving as a point of support, as a linchpin, for the most varied strategies.

In Foucault's account, four prominent groups or figures were critical in this biopolitics: hysterical women, masturbating children, the reproductive couple and the 'pervert', each of which carried significant iconic charge when deployed in subsequent discussions about 'natural' and 'normal' sexuality (Weeks 1985; Pryce 2001). For example, anti-masturbation campaigns sought to protect the child from the 'dangers' of sexual self-gratification while women were sexualized because of their importance as child-bearers, with non-reproductive and promiscuous sex deemed to cause hysteria in them. Yet most histories of sexuality have focused on the 'perverse' adult male, and the process by which same-sex activity came to define and label a specific group: the homosexual. Accused of exhibiting unnatural sexual behaviour, homosexuals became the subject of pathological analysis, being defined through medical and biological knowledges that constructed the homosexual as someone whose sexual development had deviated in some way from the 'normal' (Halperin 1986). Dominant social scripts simultaneously suggested that gay male sexuality was anti-communal and anti-social, embodying a death drive that contrasted with the future-orientation of heterosexual reproduction (Edelman 2004; Halberstam 2005).

Through specific examples of the labelling and production of specific sexual types, Foucault's analysis suggested that emerging 'common sense' ideas about sex demanded the disclosure and classification of people into different types on the basis of their sexual behaviour, something that had previously been unspoken or invisible. In each case, the intent was not necessarily the punishment of those exhibiting the 'wrong' sexualities, but the production of knowledge that encouraged sexuality of the 'right' kind. Foucault thus argued that 'sexuality exists at the point where body [the focus of the anatomo-politics of discipline] and population [the focus of regulatory biopower] meet', being a 'matter for discipline but also a matter for regularization' (Foucault *et al.* 2003, 251–2). In this sense, Foucault showed that the state's regulation of sexuality was about shaping the behaviour of both individuals and entire nations. He further suggested that identifying 'deviant' sexual types such as the homosexual or the masturbating child fulfilled a dual purpose, allowing the state to define the normal body while at the same time encouraging individuals to aspire to that norm.

Foucauldian analyses of this kind help us think about sexualities as caught up in, and moulded by, a complex range of social forces. They ultimately suggest that while our sexual identity is rooted in the corporeal materiality of the body, with its inherent capacities to feel and experience pleasure, our sexuality is created through the intersection of multiple stories (or, as Foucault would have it, *discourses*) about the sexual body (Grosz 1994). Significantly, as noted above, many of these discourses have taken shape in cities. For example, institutional state-sponsored sites such as laboratories, hospitals and clinics have been fundamental in the creation of knowledge about sex and bodies, with such institutions often profoundly embedded in the urban milieu of which they are a part. Some of these can be described as *disciplinary* spaces as they are designed for the treatment and reform of those identified as sexually deviant or perverse, preventing them from 'corrupting' others through strategies of containment and incarceration (Foucault 1977). Arguably, sexology itself took shape in such spaces, with the treatment rooms and clinics of metropolitan centres, most notably Berlin, being key to the diagnosis and classification of the diffuse sexualities of the city. The knowledge produced in these spaces was intended to help citizens adapt and orientate themselves within a sexual city whose parameters seemed to change daily (Cook 2003). The pioneering sex research of Berlin-based Magnus Hirschfeld, for instance, was clearly related to fears about the apparent mass pleasure seeking and licentious behaviour that appeared to have gripped much of the city's population (Vasudevan 2006). Berlin's pioneering role in the production and dissemination of sexual knowledge was, it should be noted, also related to the struggle of Berlin's gay identified community to develop more sympathetic understandings of homosexuality as normal rather than pathological (Stakelbeck and Frank 2003).

Beyond the psy-disciplines of psychology, psychiatry, psychoanalysis and psychotherapy, urban institutions have contributed in multiple ways to 'official' conceptualizations and understandings of sexuality. For example, the disciplines of public health, policing, law, medicine, social work and family planning all took shape in major cities. Sex was not always central to the concerns of such disciplines, though the definition of illegal, insanitary or unhealthy sex often informed their understanding of urban problems:

> Institutional actors . . . draw on a preexisting set of symbolic resources by which they articulate a specific normative understanding of sexuality and identify what constitutes a sexual problem, who or what is to blame, and how to resolve it. Health-care organizations address sexuality in terms of how sexual behavior affects personal and public health and attempt to organize the sex market to minimize the health risks of sexual activity (e.g., by promoting safe-sex campaigns). From the perspective of the police, sexual expression becomes problematic only when it threatens to upset the social order or to violate the law.
>
> (Luamann *et al.* 2004, 28)

Other non-state agencies, including voluntary organizations (e.g. religious groups, moral campaigners, health educators) have also been significant in the production of sexual knowledge, their moral discourses intersecting with 'official' medical, legal and scientific knowledge in a variety of ways as they surveyed the urban scene (Hunt 2002). In the nineteenth century, for example, locations such as railway stations, department stores and dance halls were identified by 'vigilance groups' as spaces of potentially dangerous encounter between the corrupting and corruptible (Bieri and Gerodetti 2007), while the micro-geography of city streets, urinals, parks and alleyways was transformed into a veritable theatre of sexual opportunity that needed to be surveyed and policed (Houlbrook 2000; McGhee and Moran 2000; Peniston 2001; Churchill 2004). Cartographic practices of mapping hence connected specific sexual practices to specific spaces, and suggested that the city did not have a singular sexuality, but could be divided into distinct sexual zones, including ones associated with 'dangerous sexualities' (see Case Study 1.2).

The recurring mention of Foucauldian concepts of power, knowledge and discourse in this section should underline that his work has been very influential in the study of sexualities. Moreover, although Foucault said little about space per se, his work appears particularly useful for understanding the relationships between sexuality and the city. Placed in a wider history of state control and governance, Foucault's ideas suggest that sexuality became a key domain of biopolitics because of the need to deal with the social transformation and flux associated with rapid urbanization (Slocum 2009). In this sense, identifying the location of sexual types became an important way that urban governors and agents of the state sought to make sense of this flux: discourses of sexual danger, perversity and obscenity were literally and metaphorically mapped onto specific urban sites through strategic exercises of cartography, surveying and observation. Often, such mappings were predicated on forms of environmental determinism that made connections between visible environmental conditions and diagnosable pathologies: sexual abnormality was deemed to be related to the disease and deprivation evident in particular urban locales (Reckless 1933). Once identified, these zones of 'vice' could then be subject to reactive mechanisms of surveillance, regulation, discipline and punishment designed to reform its immoral inhabitants with varying degrees of intent and success (Herdt 2009). Though arrest and incarceration of sexual miscreants was one such mechanism, the strategies of governmentality used to impose moral urban order were often more subtle, including, for example, housing policies that aimed to re-distribute populations, licensing regulations setting conditions on the operations of commercial venues and by-laws outlining appropriate conduct in public parks and streets (see Chapter Four). The importance of such strategies demonstrates that the creation of sexual order does not always rely

CASE STUDY 1.2

Venereal biopolitics: sex in Seattle

Sex has often been something that the state has taken an interest in because it appears to have public health implications. The identification of certain diseases and infections as sexually transmitted is a part of this process, with the diagnosing and treating of 'venereal' disease often involving complex forms of governance in which the state monitors the population for disease through forms of statistical epidemiology, and then seeks to discipline the groups thought to present greatest risk to society. Research shows that this has often served to stereotype particular stigmatized groups in society: for example, prostitutes, drug injectors, men who have sex with men and racial minorities were identified as high-risk groups in the transmission of HIV/AIDS in the 1980s, with subsequent public health campaigns targeting such groups given they were taken to embody a 'dangerous sexuality' (Mort 2000). This echoed earlier disputes on the causes and consequences of other venereal diseases and sexually transmitted diseases, such as gonorrhea or syphilis. As Pryce writes:

> In ascribing blame and creating *otherness*, disease has provided the basis for moral signifiers of criminality, decay, degeneracy, decadence, the construction of social standards, identities, class and racism. Such signification thereby legitimises the imperatives for social policy and moral bases for surveillance, segregation, regulation and discrimination of populations, and surveillance of the body itself. Sexually transmitted infections and the individual have been discursively reconstructed from dangerousness to risk.
>
> (Pryce 2003, 152)

In his work on VD control in mid-twentieth-century Seattle, for example, Brown (2008) suggests that there was a drive to better understand VD not just scientifically but socially. In the war years, VD was seen as a particular threat to morale in a town with a sizeable naval presence, with the large number of single, transient men and women in the city seen as too susceptible to disease if they were drawn into non-normative sex (e.g. promiscuous sex, commercial sex). Brown outlines a complex politics of education, detection and treatment in which 'responsible' citizens were encouraged to report for medical screening, and accept treatment if necessary: the unwilling could be subject to quarantine in a Treatment Centre. Fastidious records of patients were kept in terms of their place of residence, age, marital status and race, with the sexual contacts of the patient traced to construct what Brown terms *epidemiological cartographies*.

This then allowed for the identification of particular populations as being at risk, with Brown noting the way that specific bodies – those of women, 'Negros', teenagers and homosexuals in particular – were targeted in localized health promotion campaigns (see Figure 1.2). In contrast to the white, male, adult body, these were figured as promiscuous bodies, in need of both surveillance and better education. Brown concludes that Seattle's mid-twentieth-century effort to tackle the social threat of VD contained both conservative and progressive elements, driven as it was by a desire to educate and inform those who were socially disadvantaged, but also a disdain for sexual behaviour that deviated from an assumed heterosexual norm.

Such research on sexually transmitted disease accordingly suggests that the medical gaze is not one that is simply paternal or benevolent, but can be disciplinary, exercising a *biopower* that has far-reaching consequences. Clinics where sexual diseases are diagnosed are therefore not just spaces of treatment, but also locations where patients are subject to normalizing judgements, appraised in terms of risk and, ultimately, transformed into active patients who are charged with self-regulating their future sexual conduct (Pryce 2003).

Further reading: Brown (2008); Brown and Knopp (2010)

---

on the coercive or punitive policing of sexualities but the unremarkable, everyday processes of governance that organize space and time (Valverde and Cirak 2003; Hubbard *et al.* 2009).

Queering the city: destabilizing sexual orders

In any discussion of sex and the city, a spatial perspective appears essential given it encourages a focus on the strategies of urban ordering that normalize particular sexualities while marginalizing others. Perhaps the most obvious example of spatial marginalization – certainly the best documented – has been the marginalization of gay and lesbian sexualities in the city. Though there have been significant variations in the stigmatization of same-sex intimacies over time, with cities deemed more accepting than rural spaces, a wealth of scholarship on gay and lesbian life in the city suggests its presence has been fiercely contested (Higgs 1999; Peniston 2001; Aldrich 2004). Those identifying as lesbian or gay have accordingly struggled to construct spaces of autonomy within the city: many have had to chisel out surreptitious meeting spaces, exploiting the physical forms of metropolitan life to create

Figure 1.2 US Public health poster, *c.* 1940 (source: US government agency, unattributable).

meeting grounds (Beemyn 1997; Higgs 1999; Atkins 2003; Murphy and Pierce 2010). As such, numerous studies have focused on the construction of spaces where men have met other men for sex away from the gaze of the state, law and disapproving heterosexual populations, with significant literatures emerging on cottaging (Houlbrook 2000), cruising (Turner 2003) and public sex environments (Brown 2008a; Andersson 2011) (see Chapter Four). While such practices of public sex are not necessarily exclusive of lesbian identified women, they have often constructed temporary realms of male same-sex sociality, intimacy and sex in private spaces (Jennings 2007).

Knopp hence conceptualizes the 'sexual experiences of queers' as being related to the transitory way they inhabit urban space:

> Social and sexual encounters with other queers can feel safer in such contexts – on the move, passing through, inhabiting a space for a short amount of time – and a certain erotic (or just social) solidarity can, ironically, emerge from the transient and semi-anonymous nature of such experiences.
>
> (Knopp 2007, 23)

This emphasis on transitory spaces of gay identification has, however, been challenged via the emergence and proliferation of highly visible 'gay villages' in cities throughout the urban West since the 1970s (see Adler and Brenner 1992; Bouthillette 1994; Knopp 1997a), a trend taken to reflect both the power of the 'pink economy', the (sometimes grudging) public acceptance of homosexuality and the official insertion of gay spaces into city planning regimes (Miller 2005). In the latter years of the twentieth century, commercial venues in such villages became important spaces where gay identities, communities and politics were constituted, offering 'safe spaces' where those with similar predilections and sexualities could interact in relatively safe surroundings (Hindle 1994; Knopp 1998).

In many quarters, the creation of such gay spaces has been regarded as a victory on the road to full citizenship rights for sexual minorities: notable examples such as New York's Greenwich Village (see Case Study 1.3), Castro in San Francisco, Manchester Gay Village, Oxford Street in Sydney, the Marais in Paris and De Waterkant in Cape Town having been massively important symbolically, economically and politically in the lives of gay men in these respective cities (Forest 1995; Higgs 1999; Hughes 2003; Visser 2003; Sibalis 2004). Gorman-Murray and Waitt (2009) argue that the inscription of a 'gay iconography' (e.g. the rainbow flag) in the landscape of such spaces has been vitally important for gay residents (for whom it symbolizes a connection with space) as well as for non-gay residents, who acknowledge and accept

CASE STUDY 1.3

Spatializing gay identities: Greenwich Village, New York

Gay and lesbian groups were always deeply involved in the social and cultural life of New York City, with Chauncey (1995) noting that there was a sizeable and vibrant gay community in the early years of the twentieth century. This very visible community took shape around gay bathhouses, cafeterias, cabaret theatres and elegant restaurants, as well as more nefariously in the public spaces of parks, beaches and piers. There were also around a dozen 'peg houses' (gay male brothels) (Reay 2010). Yet it was New York's gay bars, cabarets and saloons that were the hub of gay life in the early twentieth century. Although speakeasies in Five Points, the Bowery, Times Square and Broadway provided spaces where streetwise working class boys ('hustlers') mixed with self-styled 'pansies' and theatrical queens, it was Greenwich Village in lower Manhattan that became the focus of downtown gay life.

Yet this relatively mainstream and integrated gay and lesbian culture was to change remarkably quickly in the 1930s as liquor laws were used to repress gay bars as part of the Depression-era 'moral backlash' against sexual 'immorality'. For Chauncey, this effectively forced homosexual culture into the 'closet', and restricted it to less visible and salubrious bars, often controlled by the mafia. The Stonewall Inn in Greenwich Village was one such establishment, a bar with a dance floor in the back room that was a popular haunt for transvestites in the 1950s and 1960s, despite alcohol ordinances in New York forbidding the sale of alcohol to the 'inappropriately gendered'. Through a combination of mafia control, pay-offs to the Sixth Precinct police and occasional tip-offs, this venue persisted until 28 June 1969, when there was an unprecedented and unexpected raid from the Federal Bureau of Alcohol Control on the pretext of exposing the mafia over-pricing and watering down of drinks. Unlike regular police raids from the Sixth Precinct, where few arrests ever took place, all present were handcuffed and some taken from the building. A contemporary news report details what happened next:

> The whole proceeding took on the aura of a homosexual Academy Awards Night. The Queens pranced out to the street blowing kisses and waving to the crowd. The crowd began to get out of hand, eyewitnesses said. Then, without warning, Queen Power exploded with all the fury of a gay atomic bomb. Queens, princesses and ladies-

> in-waiting began hurling anything they could get their polished, manicured fingernails on. Bobby pins, compacts, curlers, lipstick tubes and other femme fatale missiles were flying in the direction of the cops. The war was on. The lilies of the valley had become carnivorous jungle plants.
>
> (Lisker 1969)

Outnumbered, the federal agents barricaded themselves in the bar, calling for backup. Four police were injured in the melee. The next day, some four hundred or so turned up in the garden opposite the bar to protest against the police raid, marking the beginning of a politicized and very public protest against what was deemed homophobic policing. The Stonewall 'riots' thus marked the beginning of a putative gay rights movement in the US, with the Gay Liberation Front formed in New York only days later. Rather than acting as a homophile support network, the GLF sought to make a claim to rights based on the public recognition of homosexuality rather than mere tolerance of its private existence. Though the Stonewall Inn closed down in 1969 the first Gay Pride parade began at the gardens on Christopher Street opposite the tavern, one year on from the riots, wending its way from The Village to Central Park.

Caught up in the libratory politics of the time, and wider claims about the rights to sexual and gender equality, the gay rights movement that took shape in Greenwich Village was more militant and vocal than the homophile movements that preceded it (such as the US Mattachine Society). Though it was later to splinter, the GLF provided the impetus to the nascent gay liberation movement in the US (with groups such as Homosexuelle Aktion Westberlin in Germany and Front Homosexuel d'Action Révolutionnaire in France doing similarly in Europe). As such, the unprecedented confrontation at Stonewall marked a pivotal moment in the modern history of lesbian and gay politics. While New York's gay geographies have transformed massively since the 1960s (Colun 2009), the 'Gay Liberation' memorial in the park opposite the Stonewall Inn remains a potent reminder of New York's gay history.

Further reading: Carter (2005)

(to varying degrees) the rights of gay residences and businesses to be there. In contrast, lesbian spaces remain less likely to display the visible signifiers that mark a district as 'gay' (Podmore 2006), though studies of lesbian social networks highlight the existence of extensive neighbourhood scenes which have a 'quasi-underground character', 'enfolded in a broader countercultural milieu that does not have its own public subculture and territory'

Figure 1.3 'Gay liberation' sculpture by Georg Segal, 1991, Christopher Park, Greenwich Village (photo: author).

(Adler and Brenner 1992, 31; see also Rothenberg 1995; Atkins 2003; Nash 2006; Thomson 2007; Stella 2011).

Despite the many victories of lesbian and gay liberation movements, those lesbian, gay and bisexually identified individuals who live and work outside 'gay villages' continue to find that attempts to symbolically mark off their space (or their body) as queer may be resisted by local residents intolerant of such visible sexual difference (as Valentine 1993 relates in her account of lesbian experiences in a middle sized English town). Indeed, even when homophobic discrimination is illegal (McGhee 2004), harassment, 'gay-bashing' and even murder continue to be perpetuated by heterosexually identified individuals who regard their actions as legitimate attempts to expel or repress sexual Others (e.g. see Davis 2007 on the 'gay hate murders' around Bondi Beach, or Catungal and McCann 2010 on the murder of a gay man in Vancouver's Stanley Park). Tomsen and Markwell (2009) conclude that while hate crime against non-heterosexuals is far from ubiquitous, and possibly subsiding, it remains important in structuring the lives of those who do not conform to sexual norms. A wealth of geographical work thus shows that 'everyday' urban spaces – streets, shops, parks and so on – remain saturated with heterosexual assumptions, meaning that hiding lesbian or gay desires when in public view is a commonplace strategy to avoid unwanted attention (Namaste 1996; Kitchin 2002; Lim 2004; Blidon 2008).

For such reasons, the metaphor of the 'closet' introduced by Sedgwick (1990) has been widely used to describe both the symbolic and material geographies of lesbian and gay identified individuals, based as they are on concealment, public denial and the construction of nefarious spaces that remain invisible to those who are 'not in the know' (Brown 2000). These spaces exercise a particular form of oppression, as closeted individuals remain both known and not known, invisible to the heterosexual mainstream but understood to 'invert' their way of life. Gays and lesbians can only exit the closet, therefore, on the terms set by the mainstream, labouring under an identity imposed upon them by others who do not understand their lives because of their social and cultural erasure.

Queer theory challenges the seeming naturalness or inevitability of this situation. Often deliberately provocative, queer theory figuratively 'fucks up' established understandings of sex and sexuality. A somewhat amorphous body of work, queer theory draws sustenance from the lesbian and gay liberation movements of the 1960s and 70s, AIDS activism from the 1980s onwards, as well as more recent articulations of lesbian, gay and bisexual identity in the realms of pop, art, performance and literature (Jagose 2009). Significantly, it also draws on wider currents of post-structural thought in the social sciences that focus on the instability of language. Its relationship with feminism is sometimes fraught, but there is a fertile dialogue between feminist theory and queer theory given gender and sexuality often intersect to create gendered and sexual normativities.

Though the term queer theory was not widely used until the 1990s (being credited to Teresa de Lauretis, who used it in a 1990 conference paper), a number of earlier works (notably Foucault's) are regarded as crucial in encouraging academic reconceptualization of the relationships between bodies, sexualities and gendered performance. An early intervention here was by Adrienne Rich (1980), who forcibly argued that the effort that had been devoted by sexologists to 'explaining' homosexuality was misplaced – in her view, it was not homosexuality that needed explaining, but heterosexuality. For Rich, heterosexuality thus constituted a 'compulsory fiction' which perpetuated ideas that women were innately sexually orientated only towards men, a fiction maintained, she argued, through state practices which rewarded heterosexuals while rendering gays and lesbians invisible. Although traditionally explained with reference to biological imperatives, the dominance of heterosexuality was depicted by Rich as the product of its institutionalization in a variety of social practices, rituals and laws in a way that naturalized heterosex and foreclosed other avenues of sexual expression.

Such views were later refined by other queer theorists who were not opposed to heterosexuality per se, but the inadequacy of existing frameworks of representation for describing the sexual subject (for example, Sedgwick 1985; Fuss 1989). Influential here was Butler's notion of a 'heterosexual matrix', which describes the 'self-supporting signifying economy that wields power in the marking off of what can and cannot be thought within the terms of cultural intelligibility' (Butler 1999, 99–100). In her writing, the heterosexual matrix is described as a distinctly 'masculine sexual economy' in which gender categories support 'gender hierarchy and compulsory heterosexuality' (Butler 2000, xxviii). This heterosexual matrix enables certain identifications, foreclosing and disavowing others, with 'the repeated stylization of the body' congealing over time to produce 'the appearance of substance, a natural sort of being'. To put this more simply: by repeatedly conforming to established ideas as to how men or women are supposed to dress, talk and behave, performances of the body make certain heterosexual identities appear normal and even natural. Working with such ideas, geographers have argued that repetitive embodied performances of heterosexuality make us mistake space as *inherently* heterosexual. Furthermore, they have argued that the everyday repetition of heterosexuality becomes normalized to the extent that heterosexual space no longer appears to even be sexualized space at all (Valentine 1996).

Queer theory is hence a body of work that challenges established views of sexuality that regard it as solely biologically rooted or, conversely, merely a social construct. Central here has been the desire to disrupt and deconstruct binary oppositions – primarily homosexual/heterosexual, but also sex/gender, mind/body, self/other and so on. Queer theory has hence problematized any neat correspondence between gender, sexuality and biology, exploring the polymorphous sexual identities and practices that transgress and exceed 'normal' expectations of masculine and feminine sexuality through strategies of subversion and 'genderfuck'. The overwhelming emphasis in such work is, following Foucault, that taking up subordinate, deviant or queer sexual identities can undermine existing subject categories, dislocating the prohibitions that regulate sex:

> We must not think that by saying yes to sex, one says no to power; on the contrary, one tracks along the course laid out by the general deployment of sexuality. It is the agency of sex that we must break away from, if we aim – through a tactical reversal of the various mechanisms of sexuality – to counter the grips of power with the claims of bodies, pleasures, and knowledges, in their multiplicity and their possibility of resistance. The rallying point for the counterattack against the deployment of sexuality ought not to be sex-desire, but bodies and pleasures.
>
> (Foucault 1978, 157)

This emphasis on the embodied pleasures of sex, and the possibilities of undermining sexual order through transgressive or radical sex, has been widely debated given making a division between the radical and the normal might simply help uphold the notion of a monolithic heteronormative majority (O'Rourke 2005; Wilkinson 2010). Indeed, there are some who have suggested that gay and lesbian rights (e.g. rights to civil partnership, adoption) have only been gained through an assimilationist agenda in which the sex of homosex has been rendered less perverse and 'scary', with those having less conventional sexual tastes remaining marginalized (Richardson 2000; Duggan 2002; Weiss 2008). For some, it is the notion of sexual parody which holds more promise as a way of destabilizing heterosexuality's dominance: Butler's idea of gender masquerade (such as via drag and cross-dressing) is relevant here given it stresses all gender and sex identities are scripted, rehearsed and performed. However, it is evident that such performances can also be easily recuperated by dominant powers (see Bell *et al.* 1994; Knopp 1995; Probyn 1995); gender impersonation, for example, is often performed in ways that downplay its potential to reveal what Butler (1993) terms the *homosexual melancholia*.

Therefore, unlike certain forms of lesbian and gay politics that came before it, queer theory aims to move beyond a narrow identity politics, and seeks to destabilize any fixed identity category based upon gender or sexual orientation. As such, queer is not simply a label used to label lesbian and gay individuals, but describes a politics of transgression that is also apparent in the actions of 'queer heterosexuals' who question established sexual categories (Kitzinger and Wilkinson 1994; Schlichter 2004; O'Rourke 2005), 'heteroflexible', bi-curious, metrosexuals (Miller 2005; Frank 2008; Thompson and Morgan 2008), polyamorous individuals with multiple sexual partners (Klesse 2006; Barker and Langdridge 2010), or single, celibate and asexual individuals who refuse coupledom or sex itself (Cobb 2007; Budgeon 2008; Scherrer 2008). Work on inter-sexuality, female masculinity, male femininity and transgendered identities (Namaste 2000; Stryker and Whittle 2006; Stryker 2008b) has also been valuable in emphasizing the limitations of thinking in terms of fixed identity categories, with Doan (2004) arguing that the 'tyranny of gender' creates distinct spatialities in which 'non-binary' performances of any kind are disempowered. Such observations on the limitations of the blunt distinction sometimes made between gay/straight spaces resonate with Hemmings' (2002) work on bisexuality, which concludes that bisexually identified individuals move between gay and straight spaces, being involved in the production and consumption of both without these ever becoming bisexual spaces per se.

Describing the identities and practices of those who do not identify as lesbian or gay, or have not participated in gay liberation movements, as 'queer' troubles some: for instance, David Halperin (2003, 338) suggests there is 'much to worry about in the current hegemony of queer theory – not least that its institutionalisation, its consolidation into an academic discipline, constitutes a betrayal of its radical origins'. Against this, others have insisted that queer is a term that should not only be invoked in relation to 'authentic' studies of gay or lesbian identity politics but can critically redefine relationships between bodies, identities and cultures of all kinds (Ruffolo 2009). As Halperin (2003, 341) himself argues, 'queer is by definition whatever is at odds with the normal, the legitimate, the dominant. *There is nothing in particular to which it necessarily refers*. It is an identity without an essence.' Queer, then, has been depicted as an identity category that has no interest in consolidating or even stabilizing itself given this would itself exclude other non-queer identities. Queer in this sense signifies an identity always under construction, 'a site of permanent becoming, utopic in its negativity' curving 'endlessly towards a realization that its realization remains impossible' (Jagose 1997, 56). Given queer theory's reservations about dualistic thinking, this implies that it is important for queer analyses of urban space to avoid any assumption that urban space can be easily conceptualized as being either 'gay' or 'straight'. Following Colomina (1994, 83), it is therefore possible to argue that queering space should not involve an attempt to map gay and lesbian space, but a rethinking of the terms queer and space themselves to address 'the repressed condition of apparently stable entities, the uncanniness of everyday life'. To 'queer' a city is to hence interrogate, and destabilize, its heteronormativity.

Opening up the 'black box' of heterosexuality to explore the many possible articulations of heterosexual desire that are included or excluded within dominant constructions of heteronormativity, recent work in sexuality and space has begun to consider urban geographies of heterosexualities (in the plural). Considering the emergence of 'panic figures' and the measures used to regulate excessive, perverse or immoral forms of heterosex has been an important part of this (see Hubbard 2000). In subsequent chapters, the urban geographies of current 'panic figures', including 'sex addicts', paedophiles (always, it seems, male) and sex tourists, will be examined to suggest heteronormative ideologies cannot accommodate many expressions of male–female desire: examples of 'ladettes' and women who consume sexual entertainment will show that some women's sexual behaviours are also represented as excessive and non-normative even if these women identify as heterosexual. Geographic literatures on the disciplining of the prostitute are also useful in clarifying how the heteronormative is reproduced spatially through the regulation of commercial forms of heterosex (Chapter Two), while the overt policies of zoning and licensing which exclude lap dancing clubs and sex shops from 'family' spaces (Chapter Six) are also suggestive of the diverse geographies of heterosexuality.

This differentiation implies the need to examine the varied geographies of both hetero- and homosex – as well as those sexual practices that escape such dualistic categorization – in an effort to understand the processes which centre a particular form of monogamous, coupled heterosexuality through context-specific performances (Hubbard 2000; Brown 2008b). Crucially, these performances do not just involve overt expressions of desire or sex itself: Phillips (2006) argues that the construction of dominant heterosexuality occurs not only in the 'socio-erotic' spaces where sex occurs (e.g. the brothel, the sex club, the hotel bedroom), but also in the quiet, unobtrusive spaces of everyday practice (e.g. the city's shopping streets, office spaces, schools, parks and squares). Nast (1998) insists that such spaces are often ignored in studies of sexuality because they seem unsexy or even asexual. But the fact that such places can be saturated with innocent, unremarkable and 'natural' expressions of heterosexuality – couples holding hands, kissing in parks and so on – is of course part of the process by which a particular form of heterosex is made to appear normal and beyond question:

> Though it is valuable to study sites in which (hetero)sexuality is flaunted, it would be a mistake to see these places as representative in a broader sense of the reciprocal relationships between heterosexuality and space. Heterosexuality, sometimes visible, is commonly hidden away in . . . 'unsexy spaces'. Herein lies its power: the apparent asexuality of many different homes, workplaces, cities, landscapes and other material and metaphorical geographies conceals and naturalises the hegemonic heterosexualities that structure and dominate them, and are reproduced through them . . . In concentrating too much on overtly sexy spaces, we are letting other people and places off the hook, missing the opportunity to denaturalise heterosexuality and contest heteronormativity. By exempting apparently unsexy spaces from critical research on sexuality and space, we are tacitly accepting the self-effacement of heterosexuality as 'benign and/or asexual'.
>
> (Phillips 2006, 167)

Though many studies of urban sexuality arguably still fixate overwhelmingly on socio-erotic spaces – albeit often expunging discussions of sex itself (Brown 2008a) – there is a growing literature examining the 'benign' spaces where heteronormativity is reproduced: work on spaces of marriage (Besio *et al.* 2008), domesticity (White 2010), the family (Meah *et al.* 2008), education (Thomas 2004; Taulke-Johnson 2010) and work (Guyatt 2005) have done much to shift the focus into 'unsexy spaces'. Related to this is emerging research on 'geographies of intimacy' (Valentine 2008) which is beginning to explore the relationship between spaces of intimacy, coupling and kinship: Oswin and Olund (2010, 65) suggest that intimacy, and not sex per se, serves as the 'primary domain of the microphysics of power in modern societies' with kinship, procreation, cohabitation, family, sexual relations and love being 'matters of state as much as matters of the heart'. Many of these themes have of course been highlighted outside the literatures on sexuality and space, with Brown and Browne (2009) noting that much feminist writing on cities critiques the heteronormative assumptions that influence women's daily and intimate lives (England 1993; Wilson 1995). Bringing these ideas within the ambit of literatures on sexuality and space suggests there is much that might be gained by considering the ways that cities promote particular rituals of intimacy, coupledom and reproduction.

Conclusion

This chapter has argued that there has been an important connection between cities and sexuality since the onset of mass urbanization in the nineteenth century. However, the idea that cities are particularly important sites for the intensification of sexual desire is a contested one given rural imaginaries are also replete with representations of the countryside as a sexy and sexualized space (Bell and Valentine 1995b; Little 2003; Smith and Holt 2005; Gorman-Murray *et al.* 2008; Morgensen 2009). Despite this there appears a consensus that the excessive life and liveliness of the city has allowed a more diverse range of sexual relationships and sexual identities to be forged than has been the case in the countryside (Bech 1997; Riechl 2002; Aldrich 2004; Houlbrook 2006). As this chapter has shown, the relentlessness of urban life, the constant exposure to the gaze and physical closeness of others, the proliferation of commercial spaces of leisure and the anonymity of the crowd have combined to produce cities which present a seemingly unending flow of sexual possibilities (Bell 2001). The city has hence often been theorized, celebrated and represented as a site of sexual freedom and liberalism, a space where those with 'Other' sexualities can develop relationships, networks and communities freed from the shackles of traditional morality.

Yet while cities are sites of sexual freedom, they are also sites of power. Importantly, this power is principally that associated with the state. Key state actors involved in the production of sexual knowledge – the judiciary, police, clinicians, doctors, social workers, sex therapists and so on – tend to be most evident in the urban, with the presence of state institutions, and ever-present surveillance, seeking to discipline sexual practices in the interest of creating an ordered, efficient city. Cities are hence places where ideas about 'normal' and healthy sex are produced and disciplined. This occurs not just in those spaces deemed socio-erotic – sexualized, in effect, by sex itself (Weitman 1999) – but also those everyday spaces of intimacy, coupledom and kinship that sometimes seem far from sexy.

Some of these ideas appear to have little to do with contemporary urban theory (for example, the Marxist urbanism of David Harvey or Ed Soja), which says little about sexuality and instead sees the city as structured by capitalist processes of accumulation (see Hubbard 2006). From a Marxist perspective, the city is ordered so as to reduce class conflict while ensuring profitable production and consumption. The regulation and control of sexuality appears to have little to do with this, unless one considers the importance of *social reproduction* – the processes by which capitalist societies perpetuate themselves over generations. The policing of sex is not incidental to such processes; rather, it is integral. For capitalist cities to survive, and flourish, it has been important that they are sexually ordered, with the promotion of the monogamous, nuclear family unit, and the sexual identities it implies, appearing vital to the organization and reproduction of urban life. In turn, the spatial regulation of sexuality, intimacy and love has been integral to the reproduction of capitalism (Smith 2001). And even if the capitalist city is not what it was, with the relationship between sexuality, household formation and the urban economy becoming more fluid (see Chapter Three), the regulation of sexuality through the regulation of urban space remains important in ensuring that the city works as a site of social reproduction. Moreover, sexuality also appears increasingly important in urban economies as a commodity that can be bought and sold, with sex work (Chapter Two) and adult entertainment (Chapter Six) becoming more visible at the heart of many cities.

The remainder of this book thus seeks to develop these ideas by exploring the relationship between sex and the city across different urban landscapes, focusing on the ways that herermativity is produced and reproduced. Chapter Two, for example, considers the city's uneven sexual topography to have been exposed via studies of prostitution, past and present. This shows that the city has acted as a key marketplace for commercial sex, but rarely without limits, with the state and law seeking to impose geographical limitations on the selling of sex. Noting that many sex workers actually sell sex in their home, Chapter Three explores the sexualization of the home in greater detail, noting that dominant myths of the home as a sexually moral and ordered space do not tell the whole story: the home, it appears, can accommodate diverse sexual identities and practices, 'queer' and otherwise. In Chapter Four, the analysis turns to the sexual life of public space, which, though strongly ordered by notions of public decency, appears to offer diverse moments and spaces for sexual experimentation. While not all public sex is transgressive – far from it – this chapter shows that the legal interpretation of what is public and private upholds particular ideas of healthy and normal sexuality. Such contested ideas of what is acceptable recur in Chapter Five, where commercial spaces of night-life and leisure are scrutinized. Given these are often figured as playful sites where diverse sexual identities can be paraded, and sexual partners sought, this chapter identifies an important association between sexuality and the emergence of the leisure society. Themes of sexual commodification and the sexing of leisure recur in Chapter Six on the rise of adult entertainment, where there is an explicit emphasis on the erotic as an element of escapism from the everyday. However, the ultimate difficulty of distinguishing between spaces of leisure, pleasure and business is described in Chapter Seven, which considers the desire that flows between cities, and not just that which circulates within them. In this chapter, the centrality and importance of sex within city economies is highlighted, and the business of sex considered from a global perspective.

Ultimately, this book only can only scratch the surface of a diverse set of debates, and remains rooted in Anglophone literatures. Yet its overview of the relationship between cities and sexualities ultimately aims to show that sexuality is firmly implicated in the production and consumption of urban space. Those who design cities, as well as those who inhabit them, do so in ways that reflect their sexual predilections and prejudices, their desires and disgusts, their fears and fantasies. In the remainder of this book, this interplay will be scrutinized in more detail, focusing in particular on some of those sites where the imprints of sexuality appear sharply etched (the red light district, the lap dancing club, the nudist beach, the gay bathhouse) as well as some of those sites that, in contrast, appear somewhat 'unsexy' (the suburban estate, the shopping mall, the school and the home). Moving between questions of property (i.e. the economics and politics of real estate development) and propriety (i.e. the social and cultural norms of sexual comportment) this book interrogates these diverse spaces and modalities of urban life in an attempt not simply to show that sexuality matters in urban studies, but to think critically about the possibilities for creating more sexually liberal and egalitarian cities.

Further reading

Bell and Valentine (1995b) is often acknowledged as a seminal collection given that it mapped out a new sub-genre, namely geographies of sexualities. This edited collection contains numerous chapters that established the conceptual landscape on which subsequent studies have built.

Browne (2007) provides a review and prognosis of research on sexuality and space, and offers some useful reflections on the contributions of queer theory to the analysis of urban space.

Colomina and Bloomer (1992) is a multi-disciplinary collection focusing on the relations of architecture and sex: it was one of the first collections to have focused explicitly on the relationship of sexuality and space.

Johnston and Longhurst (2010) provide an accessible and lively introduction to geographies of sexualities, with their chapter on 'The City' a useful overview of some of the key themes in the literature.

2 The moral geographies of sex

Learning objectives

- **To understand how the city can reflect, and reproduce, dominant moral orders.**
- **To appreciate how distinctions between moral and immoral sex create 'red light districts' and tolerated spaces for commercial sex work.**
- **To be aware of the impacts of prostitution policy on the location of sex work.**

While there is a close association between sex and the city, in the last chapter it was stressed that the city is not, and never has been, a homogeneous sexual space. To the contrary, it has created the possibility of differently sexualized bodies, practices and identities co-existing in time and space. This co-existence, and the coming-together of difference, has generated multiple opportunities for sexual encounter:

> Cities have been traditionally defined on the basis of their seemingly infinite population, celebrated diversity and unconventionality, and anonymity among strangers liberated from the constraining influences of small-town social mores. For many urban denizens, the city and its never-ending flow of anonymous visitors suggests a sexualized marketplace governed by transactional relations and expectations of personal non-commitment, particularly in downtown entertainment zones where nightclubs, bars and cocktail lounges are concentrated.
>
> (Grazian 2008, 12)

The upshot is a city that is celebrated by many for its 'soft' qualities, its sexual possibilities and its utopian potential. Yet these encounters, and overlaps between different sexual lives, are also the source of anxiety and fear (Houlbrook 2006). While these sexual dangers are often said to be acutely felt by certain groups – women, sexual minorities and the elderly in particular – it is the state and law that appears most anxious to police these encounters, surveying, zoning and otherwise 'tidying up' the sexual city through acts of partition and boundary-making.

Poised between dread and delight, the sexual city is thus subject to actions intended to produce clarity. These include forms of regulation designed not so much to discipline the sexual practices of individuals but to symbolically order urban space, placing sexual bodies and identities in their 'appropriate' locations. As was emphasized in Chapter One, this is a regulation of sexuality *through* space, a form of governmentality that controls the spacing of sexuality through 'the exercise of a careful and calculated disposition of liberties and restraints' that extends across multiple disciplines such as policing, medicine and town planning (Howell *et al.* 2008, 235). Particularly important in the context of this chapter is the role of the law, which provides an important constraint on what we can do with our bodies. As legal geography suggests, the law is always spatial in the sense that it dictates the appropriateness of different actions in different spaces, and is fixated on notions of propriety, property and privacy (Blomley 2005). By making distinctions between the legality of sex acts in different spaces, the law plays a key role in constructing a veritable *moral geography*.

So what does it mean to speak of a moral geography of sexuality? Essentially, ideas of moral geography concern assumptions about what behaviour belongs in which particular places. Questions of whether certain acts blend into or transgress the character of specific urban districts are, in turn, informed by assumptions about the type of place that they are: 'high' or 'low', 'central' or 'peripheral', 'core' or 'marginal', 'public' or 'private'. When acts are considered 'in place', they evoke a sense of belonging; when out of place, they can provoke moral panic. As originally defined by cultural theorist Stanley Cohen, a moral panic is an episode in which a particular group or phenomenon becomes defined as a threat to the integrity of the nation-state. Typically prompted by an event that shocks those claiming the 'moral high ground', Herdt argues the often-visceral responses to sex acts are crucial in the definition of a particular group as a dangerous or immoral threat:

> Sexual panics in advanced welfare capitalism evoke strange, lurid, and disgusting images that merge media and popular reactions 'below the surface of civil society', targeting individuals and groups in ways that produce coherent and incoherent ideological platforms and political strategies. The conscious and unconscious resonances of this process, produce state and non-state stigma, ostracism, and social exclusion . . . and

> generate images of the monstrous. In media representations, especially, sexual panics may generate the creation of monstrous enemies – sexual scapegoats.
>
> (Herdt 2009, 2–3)

Examples here given by Herdt include the fear of the masturbation 'epidemic' that haunted the eighteenth and nineteenth centuries, moral crusades against abortion and unwed teenage mothers, anti-pornography campaigns and panics surrounding homosexuality and HIV. Though by definition these panics are expressed at a national scale (primarily through media and political discourse), most have their origins in a specific event that becomes whipped up by a hysterical media into a wider debate about the decline of national values. For example, national moral panics about paedophilia have been widespread in the post-millennial years, with instances of child and teenage sex abuse justifying a range of measures designed to curtail the threat of the paedophile (see Chapter Four on the protection of the child). The assumption that no child can consent to sexual activity with an adult means that this represents one of the great social taboos and an unanswerable case for action. A number of policies and laws have hence been promoted – from Internet filtering, not allowing gay men to work with youth organizations, sex offender re-housing, film censorship and even laws concerning zoophilia (bestiality) – on the basis that intervention is required to protect 'innocent' children from the threat of the predatory male adult, a key moral demon of our times.

This social and cultural marginalization of specific sexual types – such as the paedophile – matters in the sense not just that it creates hierarchies of moral worth, but because it informs the actions of the state and law, which respond via spatialized surveillance, regulation, discipline and punishment (Herdt 2009). In this chapter, I aim to explore this complex intersection of law, morality, sexuality and space by focusing on the urban geographies of one particular sexual 'type' – the prostitute. As a figure almost synonymous with urban life, the prostitute's presence in the city has been one fraught with contradiction: both repudiated and desired, the prostitute is known through discourses of both desire and disgust (Walkowitz 1982; White 2006). While this means that prostitution is usually accepted as an inevitable part of urban life, it has been subject to forms of legal control that confine it to specific spaces. Reviewing historical debates around the regulation of prostitution, the chapter considers the forms of spatial exclusion and incarceration that have existed under different regulatory regimes, past and present. Before doing so, however, it considers why prostitution has traditionally been characterized as a form of immoral or deviant sex, and outlines the medico-moral panics that have encouraged the state and law to criminalize many of its manifestations.

The 'whore stigma': why is prostitution seen as immoral?

In simple terms, prostitution is defined as the provision of a sexual service in return for money or payment in kind. Conventionally, this is thought to involve a female worker providing men with a quick sexual release (i.e. orgasm) through masturbation, oral, vaginal or anal sex. Other performative aspects of sex work – the adoption of particular modes of dressing, talking and acting, as well as forms of sociality between worker and client – have normally been seen as incidental to this, with the key commodity that is being bought and sold considered to be the body-as-object. The idea that this involves a debasement of Self and a reduction of the lived body to the status of an object to be bought, sold and penetrated has thus been a common trope in writing about prostitution and the sex industries more widely (witness the title of Jeffreys' 2008 treatise on commercial sex: *The Industrial Vagina*). From such perspectives, prostitution is neither 'good sex' nor 'good work': rather, it is seen as an impure form of sex, incompatible with respectable femininity.

In contrast, and especially in recent decades, sex work advocates have sought to re-conceptualize sex work as erotic labour, depicting it as no more alienating or debasing than any other form of work. However, like other forms of 'emotional work', it is clearly a form of labour that is not always easy, and may in the long term be emotionally draining given it involves the repetition of acts designed to intensify desire (see Brewis and Linstead 2000). As such, sex itself may not even be the defining feature of prostitution in many instances given sex workers are effectively enacting and selling a *performance*. Whether or not this type of performance should be equated with the emotional service labour performed by 'skilled' practitioners such as psychotherapists or child counsellors is something that Wolkowitz (2006) questions, preferring instead to draws parallels with other servicing jobs such as cleaners and nannies where dealing with bodies (and bodily fluids) is commonplace.

A number of things emerge from such discussions of sex/work, not least that sex work involves a conscious *stylization* of the body that is also present in many other forms of work, suggesting that it should not necessarily be regarded as more morally problematic or exploitative than other forms of employment. Moreover, as Zelizer (2007) notes, the relationship between client and sex worker is often not so different from the type of negotiated sex that occurs between cohabiting couples, where it might be relatively common for one partner to promise the other certain sexual services on the assumption (explicit or otherwise) that they will receive certain material rewards for doing so, whether in the form of a night out, a holiday, a new kitchen or whatever.

The obvious counter here would be that within cohabiting households, sexual relations would be more likely to be enjoyed by both parties – a dubious assertion of course, and one that denies the possibility that some sex workers also enjoy aspects of their work (Sanders 2005). For some, pleasure and prostitution are antithetical concepts, yet, as Brewis and Linstead (2000) point out, it is important to acknowledge that some sex workers report taking pleasure from their work: for some it can also be a way of seizing power rather than accepting exploitation.

Acknowledging both the agency of the prostitute and the performative power of sexualized labour is important given most representations of prostitution depict it as inevitably exploitative and debasing. Such acknowledgement is particularly valuable if one considers the diverse ways sex can be bought or sold: although prostitution has always existed in a multiplicity of forms, both indoors and outdoors, it has been widely asserted that in the late twentieth century it took on new forms and modalities as the sex industry diversified and became more accessible (Weitzer 2010). New forms of sexual commerce based on virtual exchange emerged, with telephone sex, interactive TV channels, webcams and Internet chat rooms allowing sexual services to be purchased 'at a distance' (see Hubbard and Whowell 2008). This diversification problematizes the conceptual conflation of sex work and prostitution, with the exchange of sex for payment encompassing a wider range of services and performances in which sexual relief and orgasm may be incidental or even irrelevant. Consequently, although exotic dancers, strippers, escorts, pornographic actors and telephone sex operators would not generally identify as prostitutes, all can be conceptualized as workers within a globalized sex industry (an argument considered in more detail in Chapter Seven).

So is prostitution always exploitative or demeaning? The answer is not straightforward, and depends on the specific economic and social relationships that are negotiated between worker and client(s) in different forms of prostitution. For Weitzer, this means we must reject ideas that prostitution is always exploitative or, to the contrary, always empowering:

> While exploitation and empowerment are certainly present in sex work, there is sufficient variation across time, place, and sector to demonstrate that sex work cannot be reduced to one or the other. An alternative perspective, what I call the polymorphous paradigm, holds that there is a constellation of occupational arrangements, power relations, and worker experiences. Unlike the other two perspectives, polymorphism is sensitive to complexities and to the structural conditions shaping the uneven distribution of agency, subordination, and workers' control.
>
> (Weitzer 2010, 6)

Van der Veen (2000) likewise suggests that sex work in fact comprises many class processes: for example, some workers may be involved in capitalist enterprises by working in corporate venues, while others may be self-employed, determining the terms of the prostitute/client relation. Yet others may be held in a feudal-type class relation with a pimp, or work collectively through a communal process of producing, appropriating and distributing surplus labour. It is hence impossible to make generalized claims that sex work is always exploitative. Indeed, a nuanced understanding of participation in the sex industry requires an appreciation that it has different consequences dependent on the specific relationality between bodies in different forms of prostitution. For Zelizer (2007) these relations always exist somewhere between the purely intimate and the solely monetary depending on the basis on which payment is made, meaning that there can never be an easy distinction made between the emotional/intimate and the financial/economic. In making such claims, Zelizer notes that the courts have indeed ruled that some payments and gifts given to sex workers could not be construed as payment for sex per se, and, conversely, that some people who would not consider themselves as selling sex have been doing just that by accepting gifts, food and accommodation in return for sexual favours. Questions of exploitation thus become difficult to judge given the range of possible arrangements (both spoken and unspoken) negotiated between individuals around forms of sex, love and intimacy.

Given the sheer variety of sex working, any attempt to generalize seems bound to fail. However, Bernstein (2007) suggests that identifying distinctions between different markets in sexual labour may at least provide a basis for informed discussion, with the *location* of sex work providing perhaps the most meaningful way of approaching prostitution. Indeed, in Bernstein's (2007) own work on middle class sex work it is clear that the cultivation of particular forms of 'bounded intimacy' and paid sexual engagement relies upon the production of specific spaces for sex work. In turn, the visibility, surveillance, ambience and management of these spaces shapes the nature of the relationship brokered between sex worker and client (Hubbard 1999). Most media accounts suggest the key distinction here is between street sex working and off-street work, with the former being deemed more risky and exploitative than the latter. However, much reputable social science research rejects this simplistic dichotomy, noting the diversity of sex working in cities and the movement of both clients and workers between these different spaces (see Table 2.1). Such work suggests distinctions between 'safe' indoor working and 'dangerous' street working tell only part of the story, with the safety or risk endemic to different environments strongly shaped by the intervention of the state and the law.

Table 2.1 Sites of 'direct' sex work (adapted from Sanders *et al.* 2009)

Type of market	*Characteristics*	*Geographical prominence*
Street	Visible soliciting. Use of vehicles and public spaces for sex.	Widespread despite being widely illegal under public order or prostitution acts.
Brothel (sex on premises)	Premises designed for sex work, with several workers present. Normally receptionist or 'maid' present.	Widespread despite often being illegal under brothel keeping acts. Licensed or regulated in some nations (Germany, Netherlands, parts of Australia), decriminalized in others (New Zealand).
Escort	Private workers providing outcalls (and sometimes incalls) to private hotels or homes. Often mediated through agencies and advertised though newspapers and Internet sites.	Widespread especially in Western nations and world cities with business tourism. Generally legal, though agencies and workers licensed in some jurisdictions.
Private flats (sex on premises)	Premises rented for sex work. Informal settings. Usual single worker or collective ownership.	Private business usually legal where only one worker present. Found across the world.
Homes (sex on premises)	Informal arrangement between worker and client which may be negotiated in variety of spaces.	Generally a private transaction, not formally or legally acknowledged and ubiquitous across the world.

The sheer diversity of sites where sex is bought and sold – streets, parks, massage parlours, brothels, karaoke bars, hostess clubs, hotels, homes, flats and so on (Agustin 2005) – underlines that prostitution is more diverse than it is sometimes portrayed in the media. Moreover, as recent work on male and trans sex workers shows, it not only involves women selling sex to men, but men selling to other men, men selling to women and women selling to women (Minichiello *et al.* 2003; Ashford 2008; Whowell 2010). Despite this, Weitzer suggests that when most people think of sex work, they think of the female street prostitute, a figure that has been the focus of many discussions of urban sexuality since at least the nineteenth century, as Walkowitz suggested in her groundbreaking study of Victorian prostitution:

> The prostitute was the quintessential figure of the modern urban scene. Repudiated and desired, degraded and the threatening, the prostitute attracted the attention of numerous urban male explorers.
>
> (Walkowitz 1982, 213)

Victorian representations of sex workers indeed depicted them through contradictory languages of desire and disgust. On the one hand, images of prostitutes in art and literature represented them as sexually alluring figures, the embodiment of an unfettered feminine sexuality. On the other hand, they were depicted as 'fallen' women, who, by lowering themselves to sell their bodies, had rejected the feminine virtues of chastity and homemaking. Significant outbreaks of deviance, disease or deprivation were hence mapped onto the bodies of street prostitutes, who became widely scapegoated as a cause of sexually transmitted diseases (Nead 1992). For example, William Acton, one of the most fastidious documenters of prostitution in nineteenth-century London, suggested that prostitutes could have up to twenty-five clients per day, and that one of these would inevitably be infected with venereal disease, meaning that the infected prostitute would then go on to potentially spread the disease to another 3000 men within a year. This was argued as a major threat to the defence of the nation in times of war given the armed forces provided much of the custom for prostitutes: notably, it was the prostitute herself who was figured as the key vector of disease transmission, not the soldier or sailor, who was deemed vulnerable to the advances of 'oversexed temptresses of the devil' (Himmelfarb 1995).

The idea that prostitutes were 'fallen' women whose habits had led them away from the moral and religious values inculcated during their childhood was a key motif in nineteenth-century representations of sex work. Evidentially, some sought to explain this predilection to 'vice' as pre-determined by biology rather than being associated with economic need. For example, celebrated Italian criminologists Cesare Lombroso and Ernst Sellig argued that prostitutes were effectively a sub-species, whose sexopathic and crimogenic tendencies predisposed them to their career (Evans 2003). Likewise, Tovar de Lemos (1908) concluded from his physiological studies of prostitutes that many of them had a 'male physiognomy, a tendency toward homosexuality and a low fecundity – in short, they did not tally with the standard of femininity' (cited in Evans 2003, 621). Emergent sciences of phrenology, anthropology and psychology figured the prostitute as a 'race' apart, one whose morality and behaviour was apparently written large in her physiology and appearance.

Reflecting on these pseudo-scientific accounts, as well as media discourses of the time, Corbin (1990, 25–40) identifies five dominant myths of the prostitute that emerged in nineteenth-century France: first, that she was 'putain' – 'she whose body smells bad'; second, that she was the safety valve that allowed the social body to 'excrete the excess seminal fluid that causes her stench'; third, she was associated with death, and the corpse; fourth, that she was diseased, and symbolically associated with syphilis; and, finally, that she was a submissive woman, at the beck and call of the bourgeois male. This latter theme connected to the idea that prostitution was a treacherous presence in the city, a threat to respectable society, monogamy and the family in its widest sense because of its potential to draw married men into sexual immorality. All this combined to create the 'whore stigma' – an imaginary that served to constrain the range of possible femininities deemed appropriate for respectable women.

Felski (1995, 19) consequently argues that the female street prostitute, as a figure of 'public pleasure', came to symbolize 'the abyss of a dangerous female sexuality linked to contamination, disease, and a breakdown of social hierarchies in the modern city'. By sexualizing and feminizing the public realm, the prostitute was an 'inconvenient figure' who demonstrated to male authority that its control of the city was not as complete as it would have liked others to have believed; cut loose from the bounds of monogamy, productive labour and religious asceticism, she escaped from the bounds of moralized space. While prostitutes were far from a distinct group or type, and actually came from varied class backgrounds, most coverage fixated on them as part of an urban underclass that grew up in the city's deprived neighbourhoods. Prostitution offered an escape and a distraction from poverty for such women:

> While the growth of manufacturing in the major urban centres in the late nineteenth century did open up new opportunities for women as paid workers, the options were still very limited and none of them very alluring . . . domestic service was still the most popular female occupation, offering

> women a life of confinement, hard work, low wages and the strong possibility of sexual harassment from male employers and their sons. Women who preferred the freedom of the factory had a narrow range of female jobs open to them, all of which paid around half the male rate for similar occupations. For young women living at home, this was just enough to make the effort worthwhile. For older women, or those without family or friends to supplement their wages, life was difficult to say the least. Women with dependents found it almost impossible to subsist and support their children or other relatives on factory wages, while the logistics of balancing childcare and paid work outside the home were daunting . . . Prostitution was attractive to many such women because, like outwork for the clothing industry, it offered them the chance to work from home.
>
> (Frances 1994, 33)

Working class women who sold sex, preying on the idle urban rich as much as the working man, crystallized bourgeois anxieties about public order and the moral habits of the unrespectable poor. In turn, the fact that nineteenth-century prostitution was treated as a working class, female problem allowed the middle classes to initiate campaigns to tackle the 'menace' of prostitution, absolving themselves from blame for the emergence of this social ill. Significantly, the rise of this classed politics of reform and rescue (initially led by men, but later by women – see Self 2003; Legg 2009) occurred in parallel with the introduction of governmental regulation designed to define, and control, prostitution. As this chapter will subsequently demonstrate, this control was typically exercised spatially, aiming to segregate, incarcerate and otherwise spatially 'fix' the sex worker – but not the client.

Governmentality and 'trial by space': how has prostitution been controlled in the city?

Histories of sex work suggest the nineteenth-century depiction of prostitution as immoral prompted an unprecedented attempt to discipline the sex worker (Shaver 1985). At the heart of such disciplinary systems of control were incarcerating technologies designed to simultaneously punish and reform prostitute women. Although these technologies took markedly different forms, the goal was the same: identify which women were prostitutes, round them up and lock them away in a place where the intervention of middle class reformers and religious influences could 'save them' from a life of vice. Often, such forms of incarceration were justified as measures to prevent the spread of physical diseases, especially syphilis, of which prostitutes were assumed to be a major vector of transmission.

These assumptions about medicine and morality were writ large in the work of Alexandre Parent-Duchâtelet, an obsessive moral hygienist and reformer who had been placed in charge of numerous French commissions into public health, sewerage and cholera in the early nineteenth century. His 1836 report on prostitutes in Paris was over eight years in the writing, and was heralded as a pioneering study at the time: retrospectively, it can be seen as one of the earliest sociological studies of the relations of sexuality and space. Claiming that prostitutes were as inevitable in cities as 'sewers, roads and rubbish dumps', Parent-Duchâtelet drew on both police records and his own physical examinations of prostitutes (including measurements of their labia and clitoris, their quality of voice, colour of hair and eyes) to develop a taxonomy of prostitutes. Bell (1994) notes that he produced graphs of eye and eyebrow colour for all 12,600 prostitutes registered in Paris between 1816 and 1831. Though he ultimately figured the prostitute as pathological in a socio-environmental sense rather than in a somatic or physiological manner, the net outcome of his work was to establish the prostitute as a form of pollution that needed to be controlled in the interest of social hygiene. In a Foucauldian sense, one can also interpret Parent-Duchâtelet's work as an authorized form of knowledge that had certain 'truth effects': the exhaustive cataloguing of the habits and appearance of prostitute women gave credence to Parent-Duchâtelet's claims about the need to regulate the female prostitute – and not the male client.

Writing over a century later, Corbin (1990, 4) identifies in this obsessive scrutiny of the prostitute a very real fear not of the existence of prostitution per se but the idea that prostitutes could 'come back into society', leaving their career in sex work to join the ranks of respectable society. Here, he focuses on Parent-Duchâtelet's fear of the crumbling of social categories, manifest in his assertion that 'they surround us . . . they gain access to our homes' (Parent-Duchâtelet 1836, 14). In essence, then, he stressed the importance of knowing who prostitutes were so that it would be impossible for them to 'pass on' immorality to those who did not belong to 'the class of public prostitutes' (Corbin 1990, 12). This implied a need for enclosure.

In keeping with his argument for urban prophylaxis, Parent-Duchâtelet provided the archetypal model of enclosing and surveying prostitution, a model that subsequently became enshrined as the French *regulationist* system. Chiefly associated with the Parisian Bureau des Moeurs (vice police) first established in 1802 (Harsin 1985), this model was copied by *communes* the length and breadth of France. In essence, the principle of the regulationist system was that sites of sex work needed to be 'invisible to children, honest women and even prostitutes outside the system' with enclosure 'making it possible to carry marginalisation to the limit' (Corbin 1990, 10). Ideally, this enclosed milieu was to remain constantly under the supervision of the authorities, with Corbin (1990, 9) noting that while it was 'invisible to the rest of society, it was perfectly transparent to those who supervised it'. Invoking Foucault's ideas of the power of surveillance, Corbin thus describes how this regulationist zeal resulted in strictly monitored *maisons de tolérance* (or official brothels) in designated districts (*quartiers réservés*) of most major towns. Designed to concentrate vice, the idea was that this system of regulation would purge the streets of prostitutes, but perhaps more importantly, would allow prostitutes to be surveyed, compartmentalized and hierarchized. Yet the system of enclosure needed to be policed, and the threat of prison was considered by Parent-Duchâtelet (1836) as necessary to generate a permanent terror that would inspire prostitutes to temper 'any exaggerated development of prostitution'. Hence, a system of registration was introduced requiring the voluntary (and, after 1878, compulsory) registration of prostitutes; those who refused registration suffered humiliating medical inspections and imprisonment.

This regulationist system, and the enclosure it enacted, was to prove influential in other national contexts where social reformers and commentators remained divided as to how best to deal with the Great Social Evil of the times. For some moral reformers, regulationism was tantamount to acceptance of the inevitability of vice, but for others, it offered an essentially modern and civilized solution allowing for the management of prostitution and its isolation away from the polite spaces of the city where they might corrupt the innocent:

> The Parisian *maisons de tolerance*, formerly called bordels . . . in which prostitutes are lodged gregariously, are . . . under the most complete supervision of the police. Numerous formalities must be gone through before a licence is granted by the Bureau des Moeurs, and stringent regulations must be complied with under inexorable penalties. They may not exist near places of worship, public buildings, schools, furnished hotels, or important factories. They may not be on a common staircase. They are not allowed to be near one another, within the wall, but in the Banlieue [suburbs] their concentration is imposed.
>
> (Acton 1870, 92)

For Acton and others, the French system offered a model that could be applied in other contexts. Yet the emergence of regulationism, as Howell (2009) insists, was never an abstract application of disciplinary power but a *localized* intervention in the economy of sex, and its implementation and apparatus varied according to the contingent relationship that emerged between discourses of law and practices of disciplining (see also Ogborn 1993). Yet, as Howell notes, regulationism was characteristically based on an attempt to *contain*. As such, for all the notable variations in the ways regulationism was conceived and enacted in different national contexts, Howell (2009) argues that regulationism followed the same logic: an attempt to enclose and survey the dangers of uncontrolled female prostitution lest it pollute the body politic and offend the sensibilities of polite society. The key tactics here, as encapsulated in the French system, were the identification of known prostitutes, their regular inspection and, if required, their incarceration.

Though many nations never enacted regulationism as comprehensively and extensively as was the case in France, the construction of spaces of constant surveillance was characteristic of nineteenth-century attempts to govern prostitution in the modern city. A notable example here were the Magdalene asylums that formed an integral part of many systems of municipal regulation in Britain and Ireland, albeit that they were often provided by charities and maintained a veneer of being benevolent institutions. In the UK, rescue movements such as the Church Penitentiary Association for the Reclamation of Fallen Women and the Rescue Society established homes designed to reform prostitutes through vocational training and a strictly regulated religious observation: in some cases hard labour was recommended as a means of reforming fallen women, with asylums often serving as laundry houses (the association between moral and physical cleanliness being particularly evident in such cases) (Runstedler 2006). The rhythms of the asylum were carefully controlled, with the ordering of the time-space of the Magdalene designed to instill order and morality among the inmates (Mahood 1990; Bartley 2000).

While entrance to Magdalenes was generally voluntary, other institutional spaces were places of forced incarceration, as was the case with the Lock hospitals established for the treatment of women found to be carrying venereal disease. These institutions, separate from but complementary to Magdalene asylums, became places of forced incarceration for women identified as 'common prostitutes' under the Contagious Diseases Acts of the 1860s. The goal of these Acts, the first being 1864, followed by subsequent acts in 1866, 1868 and 1869, was to prevent women from infecting male buyers of sex, syphilis being considered merely a sign of depravity for a man but a crime for women. Initially introduced in British garrison towns – and later the foreign ports in which British forces were based, including some in Australia (see Sanders 2005), Sierra Leone (Phillips 2005), South Africa (Van Heyningen 1984), India (Tambe 2009), Singapore (Levine 2003) and Hong Kong (see Case Study 2.1) – the Contagious Diseases Acts were intended to prevent the local armed forces being classified as unfit for service by virtue of having caught a venereal disease. Under surveillance from the metropolitan police, any woman suspected of being a prostitute could be subject to a (painful and humiliating) examination from a doctor: if found to be infected with venereal disease she would be sent to a Lock hospital until pronounced clean: if she refused to be examined she could be sent to prison for up to three months. In some ways, this distinction was irrelevant, as women were treated as inmates rather than patients in Lock hospitals.

The system imposed under the Contagious Diseases Acts was deemed intrusive by contemporary critics, and enacted a double standard where men could not be examined or imprisoned but the 'common prostitute' could be. Moreover, regulationism was criticized for effectively sanctioning prostitution while thereby creating a section of the population whose rights were severely curtailed when compared with other citizens. Ultimately, the Acts were repealed in 1886 in the face of popular opposition orchestrated by nascent women's groups (and, in particular, the efforts of Josephine Butler) (Self 2003). Elsewhere, variants of regulationism could be found: Sweden, for example, adopted a system of regulationism overseen by the Royal Health Committee, with a royal circular of 1812 codifying the idea that all women who worked in bars, restaurants and inns (as well as all 'loose women') should be subject to medical examination and sent to Lock hospitals (the 'kurhus') if infected (Svanström 2000).

Nineteenth-century regulationism sought to separate women into fixed categories of virtuous and fallen, often on the basis of prejudicial moral judgement: 'if the feminine ideal stood for normal, acceptable sexuality, then the prostitute represented deviant, dangerous and illicit sex' (Nead 1988, 94–5). But, tellingly, regulationism did not seek to eradicate sex work, regarding it as inevitable and even desirable in cities with large numbers of floating, unmarried men, for whom the release of sexual energies was deemed 'natural'. Some governors evidentially also saw the prostitute as having a role in saving marriages, providing men with sexual pleasures that might be denied to them within the context of their home relationship; no doubt many of the men arguing for the tighter regulation and control of prostitution were also the clients of prostitutes. Urban governors and politicians thus sought not to eradicate prostitution but to intervene in economies of sex, directing commercial sex into spaces where it could be surveyed and controlled away from the eyes of those who were most vexed about its existence.

In this respect, it was the presence of sex workers in the city's thoroughfares and public spaces that provoked most anxiety, with the provision of spaces of regulated off-street working and reformation often going hand in hand with the granting of broad discretionary powers to the police which allowed them to deal with the 'nuisance' caused by street soliciting. The 1824 Vagrancy

CASE STUDY 2.1

The colonial policing of prostitution: Hong Kong's brothel districts

Within the British colony of Hong Kong, the authorities brought in stringent controls on prostitution in the nineteenth century, most notably via the 1857 Venereal Diseases and Contagious Diseases Ordinances. These allowed for the licensing of brothels, forcible medical inspection of any prostitute selling sex to 'European' men and a strict segregation of those brothels used by Chinese and non-Chinese clients. The Register General of the island effectively became the licenser of brothels, classifying these as low class, middle class and upper class, setting different spatial limits on each, and taxing each at a rate of four Hong Kong dollars per month. Women working in brothels found to be infected were placed in hospitals for treatment, and forced to pay day fees for their treatment: brothel owners had to keep lists of everyone residing on the premises and to keep the authorities informed as to the whereabouts and health of workers. Fines for non-compliance were common and the sums generated both funded the system of regulation and paid the informants who provided the authorities with their knowledge about spaces of sex working in the colony (Levine 2003). Instances of streetwalking were also penalized, with the system of regulating brothels an attempt to render prostitution effectively invisible.

This intervention, passed a full seven years before similar Acts in the UK itself, was justified in terms of the efficiency and security of the colony, and especially the threat prostitution was deemed to pose to the army and navy in a city where white men outnumbered white women by five to one. Howell (2004) argues that the geopolitics of disease control in the colonies was built on the essentialism of race as well as sexuality, noting that the authorities were little concerned with brothels where the clients were Chinese. However, Howell suggests this intervention was not seen by the governors as a colonial imposition: they felt the Chinese authorities had previously accepted the existence of prostitution, and had therefore enacted a *de facto* regulationism. Deploying Orientalist stereotypes, the colonial governors thus depicted prostitution as a 'Chinese' problem, inevitable and natural in a city with a shortage of housing, histories of polygamy and a prevalence of servitude.

This identification of Chinese 'natives' as both racially and sexually Other led the governors to introduce what Howell (2004, 238) describes as a 'careful management of racial "others", not simply an imposition of "English" standards of "law, morality and discipline"'. This saw 'European medico-moral discourses of prostitution' giving way to a variant that highlighted the 'racial' characteristics of the Chinese rather than the 'aberrant sexuality of a small number of prostituted women' (from the urban 'underclass'). Howell argues that this emphasis on cultural difference had important outcomes for the management of colonial space:

> The geography of licensed brothels might be taken in fact as a textbook example of the segregated spatiality of colonial urbanism. The majority of brothels for 'foreign' clients were to be found in the 'Central' district of the city, along with the most important landmarks of the British and European . . . The Chinese-only brothels were located in the western district of the city of Victoria, as were the Tung Wah hospital – the premier institution of the Chinese political community – and the Lock Hospital for diseased women. Class distinctions within the non-Chinese clientele produced a distinctively managed sexual economy, a spatial order constructed through both cultural ascription and socio-political negotiation.
>
> (Howell 2004, 242)

This racially inflected regulation of prostitution confirms that exercises of biopower focused on sexual behaviour were not solely about the regulation of sexuality per se, but about the ordering of space along racial, gendered and class lines, with the irregular transactions implicit in prostitution justifying measures designed to shore up the power of minority rule in colonial Hong Kong. Today, Hong Kong remains an important centre of sex work given its world city status, with its sex markets remaining inflected along racial and national lines, with immigrant sex workers from Thailand and the Philippines mainly catering to 'Western' tourists in the hotel and night-life districts, and 'native' Chinese workers from mainland China mainly operating from one-women flats ('phoenixes') in the provinces where they principally cater to Chinese men (Kong 2006; Yang 2006).

Further reading: Levine (2003); Howell (2004)

Law in England and Wales was the first to suggest that the 'common prostitute' constituted a disorderly presence in the public streets or highways, with the 1839 Metropolitan Police Act introducing a system of fining that was taken up nationally via the 1847 Town Police Clauses Act. Such soliciting laws provided a means for the police to move prostitutes on from areas where their visible presence was deemed most offensive and provided a model that was copied elsewhere (see, for example, Frances 1994 on the concerted efforts by urban authorities to 'clean up' the streets of Australian cities).

When compared with the incarcerating technologies of regulationism, Laite (2008) argues there has been little written about the ways in which solicitation laws were deployed to impose moral order. However, the few studies that have been carried out suggest that since their inception, soliciting laws have facilitated a complaint-led policing of street sex work, with periodic police crackdowns on women working on specific streets a recurring strategy deployed in response to the complaints of local residents or business groups about prostitution's visibility in specific locales. For example, tracing the historical geographies of prostitution in Perth (Australia), McKewon (2003) notes that visible prostitution around the central city area of Northbridge in the early 1900s ultimately triggered protests from those who felt this was a blight on a family-friendly town, with this 'containment zone' ultimately closed down in 1958 when the area had transformed from being transitional space to prime real estate. Shumsky and Springer's (1981) study of the geographies of prostitution in San Francisco at the start of the twentieth century also notes that policing strategies pushed street sex working from the fringes of the city centre, with the intensification of economic activity in the central business district encouraging a more punitive policing of street working which displaced prostitution from more fashionable spaces toward the Tenderloin district that overlapped Chinatown and the 'skid row' area. Evidentially, similar processes of 'reputational segregation and racist policing' created other infamous US red light areas, including Stingaree in San Diego, Levee in Chicago and Storyville in New Orleans (see Figure 2.1) (Keire 2010).

While the vagueness of vagrancy laws gave the police considerable leeway to arrest prostitutes for wandering on the public streets and highways, in the latter half of the twentieth century such laws were largely superseded by dedicated street soliciting and, later, kerb-crawling legislation (e.g. the South African Sexual Offences Act 1957, the 1959 Street Offences Act in England and Wales; the 1960 French decree defining solicitation as debauchery and the 1985 Canadian Bill criminalizing public communication for the purposes of prostitution). Such laws seldom made prostitution itself illegal but allowed for the arrest of street working women, and, occasionally, their male 'kerb-crawling' clients, on the basis that they disturbed public sensibilities and moral order. Given the limited resource which most police forces were able (or prepared) to devote to policing prostitution, most arrests remained complaint-led, and tended to occur in phases of 'crackdown' prompted by the complaints of residents or businesses who were being exposed to the sights and sounds of street prostitution for the first time. In most cases, this served to limit

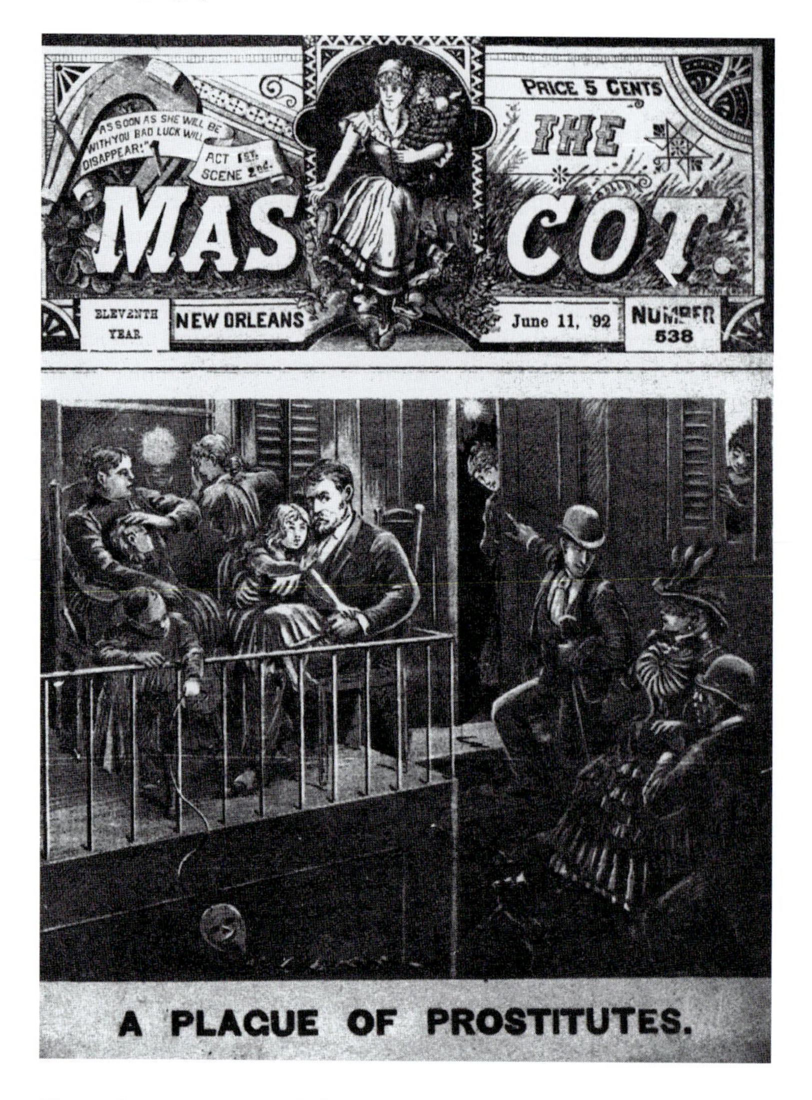

Figure 2.1 Front page of *The Mascot* newspaper, New Orleans, 11 June 1892, suggesting prostitution co-existed uneasily with home-ownership in some neighbourhoods.

possible sites of sex working to a few neighbouring streets where the presence of prostitution became well known (Hubbard 1997; Matthews 2005). Such spatial enclosure allowed the police to monitor and survey street sex markets while placing them away from areas where the presence of sex work generated most controversy and opposition (Hubbard 1997). Most sizeable towns across the urban West developed such zones, sometimes combining with other sex businesses and off-street prostitution to form a 'red light district', but more normally in isolated backstreets, where it remained invisible to the majority, and was evident principally at night (Hoigard and Finstad 1992).

So can one generalize about the control of prostitution? Based on his wide-ranging review of prostitution in the urban West, Symanski (1981) concluded that containment was the most usual 'geopolitical strategy' used in policing commercial sex:

> The specific locations of prostitution are determined by history and geopolitics: where it began and where people came to accept it; where prostitutes helped blight a neighbourhood in establishing a niche and where public opinion, financial interests and those who enforce laws have pushed prostitution or permitted it to remain.
>
> (Symanski 1981, 38)

Lowman (1992, 2009) similarly argues that the changing geographies of street prostitution in Vancouver since 1980 were a direct result of law enforcement efforts rather than the vagaries of supply and demand economics, pushing it towards less valued spaces in response to the complaints of business owners and landlords (see Case Study 2.2). McKeganey and Barnard's (1996) ethnographic study of street workers in Glasgow also suggests that the identification of a permissible working area was at the heart of the everyday negotiations between police and prostitutes, with workers who transgressed beyond boundaries deemed acceptable by the police undoubtedly charged if caught. In Birmingham too, Hubbard and Sanders (2003) showed that the location of street sex work has historically shifted in response to police action, echoing the view that the policing of prostitution *creates* red light districts by effectively making it impossible for clients and prostitutes to meet elsewhere: the state and law, in effect, has been productive of the spaces of vice whose continued existence then justifies their intervention.

The policing of street sex markets can thus be viewed as part of a continuing (but contested) process involving the separation of disorderly prostitution from orderly and 'respectable' sexuality, removing sex workers from areas where their manner of dressing and soliciting marks them as a 'polluting' presence. On the basis of this, it could be argued that the location of street prostitution reflects urban property values, with prostitution pushed away from the wealthiest residential and business areas (Larsen 1992). This given, the toleration of street sex working in transitional and industrial zones has sometimes been formalized in the interests of protecting residential and business areas which might be adversely effected by the secondary impacts and *negative externalities* of sex work. In the Netherlands, for example, 'tippelzones' were

CASE STUDY 2.2

The dangers of street sex work: Vancouver's missing women

As a centre for forestry, canning, milling and transportation, Vancouver has always had a substantial community of sex working women catering to both local workers and those passing through on business. This resulted in the emergence of notable brothel districts from the 1880s in Canton and Shanghai Streets in Chinatown, as well as West Hastings and Mount Pleasant. The identification and tacit recognition of these 'restricted districts' allowed the police to focus their efforts on eradicating streetwalking: however, in later decades coalescing concerns about disease and immorality encouraged a more intrusive surveillance of the brothels:

> Throughout the first half of the 20th century . . . official policy concerning prostitution vacillated between controlled toleration and moral zeal. During times of abolitionist fervour, sex workers were flushed out of brothels, hustled off [the] streets, charged as inmates or vagrants, fined, occasionally jailed, then discharged only to become vulnerable, once again, to waves of civic protest.
>
> (Ross 2010, 198)

Ross proceeds to document how patterns of sex work transformed dramatically in the wake of the 1975 closure of many of the West End clubs (including the infamous Penthouse Club) where prostitutes plied their trade, partly at the instigation of gay entrepreneurs and business owners who were gentrifying the area. Ross describes how a heterogeneous sex working community (including numerous transsexual and male prostitutes) took shape on the streets of the West End, largely free from the control of pimps and procurers, but opposed by community organizations that sought to remove 'sex deviance'. Stressing that sex work blighted the West End, CROWE (Concerned Residents of the West End) began picketing the streets in an effort to 'Shame the Johns' [clients] and 'reclaim' the streets for the residential community: in 1982 the city passed anti-soliciting by-laws that forbade the buying or selling of sex on Vancouver's streets. Backed with punitive fines, this broke up the sex working community and pushed prostitution back towards its traditional location in Chinatown and the Downtown Eastside.

Though the West End remained an important beat late into the 1980s, especially the 'high track' stroll of a six-block commercial area on Richards and Seymour (Lowman 2009), the displacement of much street work to the Downtown Eastside following the anti-communication legislation of 1985, and especially its toleration in the poorly lit alleyways north of Hastings Street, was a factor in an increasing number of attacks on prostitutes. Before the passing of the anti-soliciting act in 1985, there had been one murder of a prostitute per year; afterwards it rose to more than seven per year, and over sixty in total in the next decade. The disappearance and unexplained murder of sex working women in Vancouver – its 'missing women' – exposed the failures of prostitution regulation to create safer working conditions for sex workers: the fact many of these were indigenous, native and non-white workers helped the media weave this into a panic about a serial murderer ('ripper') preying on the vulnerable rather than questioning the logic of prostitution policy (Pitman 2002). The subsequent arrest of Robert Pickton, believed to be responsible for at least twenty-six murders, consolidated this 'ripper discourse', meaning that there was less discussion of the vulnerabilities – or police incompetence – surrounding street sex work in Vancouver.

For Ross, the process by which prostitutes in Vancouver have been discredited and dehumanized involved representations that depicted streetwalkers as sources of both moral and physical contagion. Lowman (2009) concurs, arguing that the media discourse was dominated by calls to get rid of prostitution, creating a social milieu in which violence against sex workers could flourish. An annual Women's Memorial March, which begins at the corner of Hastings and Main Street in the Downtown Eastside, continues to draw attention to the large number of unsolved sex worker murders, something that appeared to influence the Ontario Supreme Court's decision to overturn the Canadian laws prohibiting soliciting in 2010.

Further reading: Ross (2010)

---

established in the 1970s and 1980s where car-borne clients could solicit workers, taking them to nearby car parking spaces designed with modesty screens, discrete surveillance and bins for the disposal of used condoms (Van Doorninck and Campbell 2006). Living rooms ('karmer') alongside the tippelzones provided a space for workers to shelter, get health and legal advice from social workers and socialize with one another (Wagenaar and Altink 2009). Noting the success of these initiatives in some Dutch cities, this model has been copied elsewhere. In the UK, for example, 'managed zones' were established in the 1990s in Preston, Northampton, Bolton and Edinburgh, allowing sex workers to work at designated times without fear of arrest so long as they followed certain codes of behaviour (Matthews 2005). In Manchester, male and female sex workers were able to access sexual health, drug-use and legal support from outreach workers who patrolled the unofficial 'tolerance zone'. However, it was Liverpool that became best known for pursuing an overt policy of zoning. The proposed zone followed the murder of two sex workers in July 2003, with local surveys suggesting that sex workers, residents and businesses were in favour of the plans (Bellis *et al.* 2007), and research concluding that zones offered supportive surveillance by outreach and medical services as well as punitive surveillance by police:

> Whilst it is acknowledged that managed zones are in no way problem free and could also potentially be used in order to push a morality agenda (especially if they include mandatory testing for sexually transmitted infections or an undue emphasis is placed on gaining legitimate employment), they have the potential to create a safe space. A safe space in which ethical relationships which recognise the vulnerability of human life can emerge. A space for asking the other what is needed to make her life more bearable, as opposed to presupposing what the other needs.
>
> (Carline 2010, 53)

However, against advice from many working in the field, managed zones for sex workers were officially rejected by the Home Office (2006, 51), whose Coordinated Prostitution Policy asserted there is no place for street work in 'civilised societies'. This ruling stymied Liverpool's efforts to establish a managed zone, despite its attempts to depict its proposal as part of a tradition of municipal improvement and civic renewal as it approached the 2008 celebrations that heralded its tenure as European City of Culture (Howell *et al.* 2008).

Given their lack of official recognition, zones of tolerance and areas of street prostitution have been easily sacrificed when transitional and non-residential areas hosting street sex work have been targeted for redevelopment. This was the case in Leith Docks, Edinburgh, where there was an unofficial toleration zone negotiated via an informal agreement between police, prostitutes and local health projects on the basis that no more than twenty women should work the zone at any one time and that no drug dealing would be tolerated. Though the zone operated well for more than a decade, the Docks area underwent considerable development in the late 1990s, and the incoming residents began to complain about nuisance and noise associated with street soliciting (Bondi 1997). Despite efforts to relocate the zone, it closed in 2001 in the face of public pressure: subsequently, street sex working has been widely dispersed across the city, making the work of health projects difficult (Hubbard *et al.* 2007).

The displacement of prostitution by redevelopment is a widely noted phenomenon across the urban West (e.g. see Kerkin 2004 on St Kilda, Melbourne; Löw and Ruhne 2009 on Frankfurt; and Barber 2010 on Amsterdam). This suggests that there are important connections between sexuality and urban processes like *gentrification* that are normally understood as primarily about class conflict rather than sex. Indeed, displacement of established areas of street sex work appears particularly pronounced in the context of what Neil Smith (2002) describes as 'third-wave' gentrification, which is led by corporate investors keen to draw back affluent consumers to the central city. This is not to say the residents attracted 'back' to the city are drawn exclusively from the ranks of the mobile and affluent, but the target market tends to be young individuals, couples and families who have the purchasing power necessary to buy into metropolitan living (Zukin 1995) (see also Chapter Three on city centre living). As Smith notes, these gentrifiers characteristically identify themselves as being 'streetwise', and claim to be attracted to the city centre because it offers a contrast to the mundane nature of suburban living. Yet, simultaneously, this population seems remarkably anxious about individuals whom they regard as an un-aesthetic presence in 'their' urban space (Binnie *et al.* 2006). This implies a close connection between urban gentrification and urban *revanchism* – the process whereby the gentrifying middle classes seek to exclude those users of urban space who trouble them. Hence, sex workers can be the target of exclusionary actions and rhetoric, alongside a wide range of 'minority' groups including buskers, skateboarders, the homeless, beggars, leafleters, teenagers and street entertainers, all of whom are regarded as anti-social presences in the consumer-fixated gentrified city centre. As Sanders (2009a, 511) argues, by being labelled as 'anti sexual' – unclean, unwanted and a symbol of decay – 'the iconic whore figure is entirely out of step with the gentrified notions of modern city living and leisure spaces' and becomes 'an easy target for removal'.

Hiding 'vice' in the landscape: what are the contemporary geographies of sex work?

This chapter has suggested that sex work cannot easily be 'designed out' via police crackdowns or environmental interventions: the evident appetite for sexual services in large cities means that there is always supply and demand (see Hubbard 1999). The intractability of sex working means that, in most nations, an *abolitionist* stance remains to the fore, with prostitution legal but brothel management, procuring and pimping criminalized to prevent third parties benefitting from, or exploiting, the sex work of others. In practical terms, this means it is difficult to sell sex unless it is a consensual transaction between two adults conducted in private, with no involvement of a third party (whether manager, pimp or partner). Notable exceptions include the Netherlands, Germany, New South Wales and Victoria (Australia), and Nevada (US) where state-licensed brothels allow women (and, very occasionally, men) to sell sex, subject to workers having appropriate work permits and being subject to medical screening:

> Legal brothel prostitution is justified by a discourse that maintains that (1) the sale of sex is one of the world's oldest professions and is unlikely to disappear; (2) state and local government-regulated prostitution is superior to illegal prostitution insofar as it is allows for limitations on what is sold, on the terms and conditions of sales, and on brothel ownership and employment practices; (3) such businesses are revenue generating; (4) legal prostitution provides a valuable service to certain individuals who have desires that cannot be easily fulfilled otherwise; and (5) limiting such activities to particular licensed venues curtails related criminal activities (drugs, pimping, violence) and helps control the spread of disease.
>
> (Hausbeck and Brents 2010, 171)

Licensing premises, rather than workers, can theoretically make it easier for the state to collect revenue from prostitution, an activity notoriously hard to tax (Harcourt *et al.* 2005). Licensing can also place considerable restriction on the location, visibility and opening hours of such premises. For example, Amsterdam's famous 'windows' are only licensed in three neighbourhoods (Wallen, de Pijp and Singel) while in Germany there are innumerable 'Restricted area decrees' which ban brothels in residential areas (Löw and Ruhne 2009). It can also allow for the imposition of minimal environmental standards, including the installation of CCTV monitoring and 'panic buttons' that provide some security for working women (see Figure 2.2).

Despite recognition of sex work as a formal land use, planning and licensing processes rarely regard brothels as equivalent to other commercial or retail businesses, and often treat these as noxious or undesirable land uses. Even in New Zealand, whose 2003 Prostitution Reform Act decriminalized all forms of sex working, local authorities enact by-laws to control the visibility and location of brothels, with Brothel Operator Certification conditional on meeting certain standards. Such licensing allows a degree of spatial control to be exercised and for the authorities to place sex work where it will have few 'negative externalities' or secondary effects. This is often based on assumptions about the predilections and vulnerabilities of particular community groups, with policies tending to favour brothel locations away from schools, religious facilities and 'family areas', in industrial areas or simply

Figure 2.2 A safe working environment? Brothel interior, Sydney (photo: Great Roe/Creative Commons).

outside the city limits. A somewhat extreme example of this is provided by the Zona Galactica, located some four miles outside Tuxtla Gutierrez, Mexico (Kelly 2008). This is a municipally licensed brothel established in 1991 as part of an effort to 'cleanse' the city, contains 180 rooms rented from landlords alongside a clinic, school and a small jail staffed by police. All workers in the Zone have weekly health checks and must carry a Sanitary Control Card. 'Clandestine' prostitution practised outside the Zone has been subject to police raids on the pretext of registering workers for health purposes. Kelly (2008) describes the establishment of the Zone away from the city as a remarkable attempt to promote 'social hygiene', and one that the authorities regard as modern, progressive and ordered, yet also notes that the zone may not be a very safe space in which to sell sex given that the competitive nature of the environment discourages workers from insisting on safe sex practice.

In contrast, studies across UK cities have suggested that indoor environments where women can work together are safer than street soliciting (Sanders and Campbell 2007). In spite of this, and some positive evaluations of the consequences of decriminalizing or legalizing brothels elsewhere in the world (Albert and Warner 1995; Brents and Hausbeck 2005; Groves *et al.* 2008;

CASE STUDY 2.3

Regulated sex work: licensed brothels in the city of Sydney, New South Wales

Traditionally, most Australian states have pursued strategies of containment for sex work, tolerating it only under certain circumstances and generally prohibiting street sex work. Sullivan (2010) nonetheless describes a 'moderate opening' in the space for legal sex work in all states since the 1980s, which she argues is the combined result of feminist movements arguing for legal change, the strength of sex worker advocacy groups and a liberal authoritarianism that increasingly regards sex workers and their clients as able to take some responsibility for self-governance.

In Australia this is evident in varied forms of decriminalization and legalization across the different states, generating the possibility that planning – and related bodies of licensing and environmental law – can have a say in where (and how) sex work occurs. For example, the City of Sydney has produced a comprehensive set of guidelines (the Adult Entertainment and Sex Industry Premises Development Control Plan, 2006) recognizing and regulating the location, design and operation of sex industry premises through the provision of comprehensive planning controls.

In Sydney, many inner city areas – Kings Cross, Woolloomooloo, Darlinghurst – have been associated with prostitution since the nineteenth century, being the focus of the city's on-street sex markets. The growth of other sex industry premises in the latter half of the twentieth century (massage parlours, brothels and gay bathhouses) nonetheless created some tensions in these areas, especially when advertised garishly and through the activities of the 'spruikers' who lined the pavements of Kings Cross in the evening, hawking for trade. Though off-street work was technically illegal under the Restricted Premises Act 1943, gradual political pressure and the decriminalization of sex working premises in New South Wales in 1979 led to the 1995 Disorderly Houses Amendment Act which effectively repealed those parts of the 1943 act that made it a criminal act to operate a sex-on-premises venue in the state.

This repeal effectively removed the prohibition on private premises being used for the purposes of prostitution, deeming that general planning and environmental health legislation were sufficient to regulate sex work (as contrasted with the *legalization* that occurs in some parts of the world that demands registration of premises). While New South Wales councils cannot ban brothels outright, planning controls have however been used to restrict brothels to industrial locations only: under s.17(5)(a) of the Disorderly Houses Amendment Act, 1995 a brothel can be closed if it is operating 'near or within view from a church, hospital, school or any place regularly frequented by children for recreational or cultural activities' or if it is causing a disturbance by 'reason of noise or vehicular traffic' (s. 17(5)(b)) (Brown *et al.* 2006).

However, Crofts (2007) suggests that the failure of the state to provide spaces for sex work outside industrial areas leaves many of those choosing to work from their own home in a somewhat difficult situation. These businesses are notably different from commercial sex services premises: private sex workers generally see fewer clients than sex workers in commercial sex services premises, and the time private workers spend with clients is usually flexible. Given around 40 per cent of all sex workers in the city work from such premises, and that the type of property available in areas designated for sex work are quite unsuitable or too expensive for such sex workers, it appears that the current planning guidelines are unable to adequately accommodate the range of sex working that is undertaken in the city.

Sullivan (2010) suggests that in comparison with the situation in other Australian states, sex workers in New South Wales are in a privileged position and able to pursue a variety of ways of working, whether for employers or as part of a collective. Recognizing sex premises as legitimate land uses also allows public deliberation over the suitability of sex work in different communities. Yet for all this, the seemingly enlightened approach taken by NSW boils down to a relatively simple aim: to limit sex premises to the areas 'where they are least likely to offend' (*Hornsby v Martyn*, New South Wales Land and Environment Court, 2004). In this sense, the regulation of the sex industry in Sydney still reinforces the notion that sex businesses are *disorderly* (see Crofts 2007).

Further reading: Crofts (2007); Prior (2008)

Abel *et al.* 2009), the British government has continued to stress that its aim of 'disrupting sex markets' applies equally to both indoor and outdoor environments. Significantly, the 2009 Crime and Policing Act introduced a range of new offences relating to prostitution, with the police able to enforce a brothel closure if 'an officer has reasonable grounds for believing that the making of a closure order . . . is necessary to prevent the premises being used for activities related to one or more specified prostitution or pornography offences'. Quite why these powers have been ushered in at a time when there is increased recognition of prostitution as a form of work can be explained with reference to the emerging discourse that sex work encourages exploitation and sex trafficking (Day 2009). Establishing the idea that reducing demand is the best way of tackling exploitation, policy-makers in the UK have argued repeatedly that clients are uniformly sexist, uncaring and ignorant, and are uninterested in whether the women they are buying sex from are forced to do so or not (Sanders 2008a, 2009b). Given that there are already laws concerning forced sex and rape, the introduction of a strict liability offence in 2009 for buying sex from a prostitute exploited for gain is just the latest in a series of symbolic laws which regard sex work as gendered exploitation (Brooks-Gordon 2010; Carline 2011).

What is clear from this is that prostitution policy remains predicated on the idea that prostitution always and inevitably involves a female worker selling sex to a heterosexually identified man. Much of the academic literature also remains fixated upon this form of prostitution. However, there are many other possible configurations of sex work. For example, while male sex work is relatively invisible when compared with sometimes spectacularized spaces of female sex working (Shah 2006), research from a variety of nations indicates that male and trans sex work is widespread, variously associated with public sex environments (see Chapter Five), gay bars, bathhouses and saunas as well as private homes (Aggleton 1999; Gaffney and Beverley 2001; Hall 2007; Leary and Minichiello 2007; Smith *et al.* 2008; Kong 2009; Browne *et al.* 2010; Ozbay 2010; Reay 2010). This suggests that studies of sex work, with their disproportionate focus on certain spaces (street environments) and particular actors (female prostitutes/male clients) need to be complemented by analyses considering the different sexualities and genders of both sex workers and clients, as well as the increasing overlap between real and virtual spaces of sex working (Hubbard and Whowell 2008).

Conclusion

This chapter has shown that prostitution has long been part of the urban scene, but its presence has always been contested because of the moral anxieties it provokes. Sex workers have rarely been free to ply their trade where they wish, with different strategies of state-sponsored governmentality and surveillance having been used to contain, exclude or control. From the perspective of the state, such regulation fulfils dual purposes. First and foremost, it reduces the visibility of sex work for those who might be offended by the sights and sounds of commercial sex or do not wish to participate. In most jurisdictions there has been a clear anxiety to make such spaces off-limits to children, as well as specific cultural and religious groups who object to commercial sex on moral grounds. More generally, this anxiety to distance sex work from 'sensitive' populations has extended to encompass the protection of 'family' spaces, with exclusionary actions distancing sex work from those areas where monogamous heterosex is assumed to be dominant. Through strategies of spatial containment, the threat of moral contagion is apparently curtailed.

Second, containing sex work in specific spaces allows the forces of law and order to ensure that these spaces provide controlled environments for commercial sex, via an enclosure of the bodies of sex workers and their clients. Surveillance of such spaces by state agents (e.g. police, the judiciary, doctors and social workers) constitutes a form of biopower given this allows a level of monitoring and control to be exercised through a focus on sexualized bodies. Perhaps most evident under the Contagious Diseases Acts of the nineteenth century, the spatial enclosure of sex work has latterly confined it to commercial spaces where participants are subject to forms of surveillance via CCTV, licensing and outreach projects rather than the 'punitive' gaze of the state and the law. Nevertheless, at the heart of such forms of governmentality is a desire to discipline commercial sex and contain it in spaces where it remains invisible to the majority, but not beyond the reach of the state and the law (Hubbard 1999).

This said, recent attempts to bring prostitution within the ambit of licensing and planning law suggest an important shift in the regulation of prostitution from coercive control via criminal law to a more diffuse social control:

> Recent regulation operates by reference to the control of the space in which the activity of prostitution takes place, and by reference to community public health and amenity standards. In effect, it requires the disciplining and self-disciplining of the 'body' within the spatial and public health parameters defined by the bureaucratic state. In this transformation of regulatory control, the instrumental 'technology' of the law has shifted from an embodiment of the moral force of the sovereign state via a criminal law . . . to a statutory framework that seeks to make safe: to contain the 'vice' by imposing on the body a system of constraints and privations, obligations and prohibitions.
>
> (Godden 2001, 78)

Though intended to improve public health and prevent crime, prostitution policy is still accused of continuing to push sex workers into spaces that systematically expose them to violence, with Kinnell (2008, 65) arguing 'the legal framework makes all forms of sex work more dangerous'. Sex worker advocates thus make the case that prostitution should be regarded as part of the normal service economy, with sex-related businesses recognized as *recreational* settings where consenting adults are free to purchase sexual services. Bernstein (2007, 127) persuasively develops this argument, suggesting that commercial sex can be easily incorporated within the 'normalised field of commercial practices' given the separation of sex and reproduction (see also Bauman 2003). For Bernstein (2007), this is reflected in the changing nature of the commercial sex transaction, with the search for 'quick release' apparently being superseded by clients' search for a more 'leisurely' form of intimate encounter in which sex itself is only part of what is being bought and sold (see, for example, Zheng 2009 on the importance of singing, dancing and flirting in the karaoke and banqueting bars of post-socialist China). This collision of sex and leisure is one that is making it increasingly hard to say where the spaces of sex work begin and end – a theme picked up in Chapter Six's exploration of the geographies of 'adult entertainment'.

Further reading

Hubbard (1999) provides a fuller summary of the campaigns of community opposition reviewed in this chapter, and includes a more fulsome consideration of the historical geographies of sex working.

Sanders *et al.* (2009) is an accessible text considering the contemporary sociologies of sex working. While UK focused, it draws on a wide and international range of pro-sex worker literatures to challenge many myths of the exploited, criminal and anti-social sex worker.

Symanski (1981) was perhaps the first book-length treatment of prostitution by a geographer. While Symanski's research was controversial in many respects, this book flags up many of the issues about morality and the law that are explored in this chapter.

3 Domesticating sex

Learning objectives

- To appreciate how the equation between sexuality, monogamy and 'the family' is reflected in the design and appearance of residential landscapes.
- To understand the consequences of this for individuals with other lifestyles and sexual identifications.
- To consider the possibilities for household arrangements and relations to promote non-monogamous and non-normative sexualities.

While red light districts and zones of street prostitution are often understood to be the city's 'sex zones' (Cameron 2004), this is misleading given sex itself happens in a variety of urban spaces, and not just the streets, parks, flats, saunas and brothels where sex is bought and sold. Yet the notion of a 'city of zones' has been an important one in the histories of planning and design in the urban West. Since the inception of town and country planning, dividing the city into distinctive and legible zones has been one of the key ways that planners and urban governors have sought to impart order on the metropolis, with plans typically making distinctions between retail cores, industrial areas, business districts, residential suburbs and so on. In the planner's imagination sex is thought to properly belong in residential spaces where it is safely *domesticated* in the context of reproductive, monogamous relations. In turn, this imagination feeds on stereotyped views of gender and sexual roles (Hooper 1998), with numerous feminist critiques of modern planning pointing out that the spatial ordering imparted through urban planning has been based on particular ideas about the desirability of separating production and reproduction (McDowell 1983). Classic models of planning assumed that the male was the breadwinner, working in the city proper in business hours, while the woman was charged with social reproduction: preparing meals, doing the shopping and caring for children. The home was central to this spatial imagination, being the hub of traditional family life and, by implication, the site where monogamous heterosexuality was institutionalized and normalized.

Such processes of spatial ordering have been suggested to be central to the making of 'man-made cities' (Hayden 1980; Roberts 1991; Booth *et al.* 1996). Moreover, by setting aside space for housing, and identifying this as the right and proper space for social reproduction, processes of planning and zoning have also demarcated a private sphere of 'embodied' activity away from the 'disembodied' public sphere of political life (Duncan 1996). This implies home is the appropriate place for the washing, dressing and resting of the body, as well as the intimate performances of sex itself (Valentine 2008). But conversely homes are often presented as thoroughly de-sexed, as if sex itself does not happen there: if it does, it is imagined to be restricted to the 'master bedroom', that private space within the home that is out of the gaze of neighbours, visitors or children. This creates a taken-for-granted assumption that domestic sex is ordered (and moral) because it is not 'on display'. In fact, as this chapter shows, if one pulls back the net curtains of the 'average' suburban home one can reveal promiscuity, perversion and parody aplenty. Suburbia can be sexy too.

The mismatch between the imagery of domestic, privatized sex and its lived existence is one of the key themes in an emergent queer reading of the home which demonstrates that homespaces can be subversive sites and not simply ones subject to the dominant logics of heteronormativity. It is also a theme in work on the heterosexual home, with studies having shown how home spaces can be entwined in practices of 'wife-swapping' (Denfeld and Gordon 1970), swinging (Jenks 1998), the sale of sex toys (Storr 2002) and the production of pornography (Paasonen 2010). But this duplicity is also one that has been exposed in legions of British films and sitcoms (from the ribald humour of the *Carry On* series through to the soft porn of *Confessions of a Window Cleaner*) that mock the sexual morality of the suburban 'gnome zone' (Medhurst 1997). More celebrated films like *American Beauty* or *Blue Velvet*, and novels like *The Bhuddha of Suburbia* also emphasize the way homespaces can vacillate between sexual normality and queerness. More recently, *Desperate Housewives*' portrayal of the complex love lives of women living in the neatly manicured suburban landscapes of Wisteria Lane also suggests that the equation of suburban life and monogamous heterosexuality is something of a myth. But the existence of gay suburbanites, queer homes or sex parties in suburbia has not been enough to undermine the persistent myth of a separation between a sometimes vulgar public city and the polite (hetero)-sexuality that proceeds behind closed doors. The persistence of this myth is a key theme explored in this chapter (and the following chapter on public sex) given that the ideal of separate public and private realms continues to have important impacts on our sexual lives.

This chapter accordingly works through a series of debates concerning 'homemaking' – a set of social and gender relations in which individuals are supposed to take up particular sexual roles. Noting how these gender/sex identities are manifest in housing design (and the notion of the 'nuclear household'), the chapter explores the assumptions about (hetero)sexuality that are writ large in residential landscapes, as well as detailing the 'family fantasies' that are maintained through legal and policy measures designed to maintain domestic order (such as the spatial exclusion of sex offenders or the designation of prostitute-free zones in 'family' areas). Ultimately, the chapter seeks to queer dominant representations of domestic sex and love, asking whether other household forms are possible or desirable.

Heterosexual homes: in what ways does housing promote the nuclear family?

The idea of home has many different connotations, and varies across different cultures. However, the dominant representation of home enshrined in Western media, popular culture and governmental policies tends to be that of a private space in which members of a 'nuclear' heterosexual family develop intimacies and senses of belonging between them (Blunt and Downing 2006). In common usage, the notion of the nuclear family describes a family unit consisting of a father, a mother and their children living within a single household. This unit is defined as nuclear because it is thought to be able to exercise a level of independence in terms of economic and social reproduction. As was noted in Chapter One, it is the social unit around which contemporary urban life has been organized and developed. The fact that globally this type of household is in the minority (as compared with single households, shared households and households containing extended families) does not alter the privileging of this family formation, nor its idealization as the preferred arrangement in which to develop forms of love and companionship. More than just a description of a household type, the nuclear family provides the 'common-sense' model of how we should live and behave, materializing a series of daily, weekly and annual routines which institutionalize future-orientated forms of reproduction:

> The time of reproduction is ruled by a biological clock for women and by strict bourgeois rules of respectability and scheduling for married couples.

> Obviously, not all people who have children keep or even are able to keep reproductive time, but many and possibly most people believe that the scheduling of repro-time is natural and desirable. Family time refers to the normative scheduling of daily life (early to bed, early to rise) that accompanies the practice of child rearing. This timetable is governed by an imagined set of children's needs, and it relates to beliefs about children's health and healthful environments for child rearing. The time of inheritance refers to an overview of generational time within which values, wealth, goods, and morals are passed through family ties from one generation to the next. It also connects the family to the historical past of the nation, and glances ahead to connect the family to the future of both familial and national stability. In this category we can include the kinds of hypothetical temporality – the time of 'what if' – that demands protection in the way of insurance policies, health care, and wills.
>
> (Halberstam 2005, 13)

The queer perspective offered by Halberstam encourages us to explore how this organization of time and space privileges heteronormativity. It also asks us to consider whether alternative, queer organizations of community, sexual embodiment and identity might be possible or desirable.

Yet while the nuclear family appears to have been integral to the spatial organization of modern, Western cities, some historians have suggested it has a longer precedent. Indeed, some studies suggest that the model of a cohabiting couple living with children was the norm in pre-medieval times (Hareven 1976), while imperial projects of secular nation building from the eighteenth century onwards were clearly predicated on the idealization of the married, nuclear family (Nast 1998). But there is little dispute that housing, the household and the nuclear family became most closely aligned in the modern post-war era, when homemaking became integral to national projects of modern citizen-making and re-population (Ross 1996). Programmes of redevelopment in the 1950s and 1960s (for example, the UK New Towns movement) enshrined the ideal of the nuclear household through the planning of residential housing units based around certain key neighbourhood facilities such as schools, health centres and shops (all of which were strongly feminized spaces). While women were expected to remain within the neighbourhood unit, tending to children and home, men were anticipated to take roles as workers in the city, with transport networks privileging the daily movements of male workers out of, and back to, their homes (Hubbard and Lilley 2004). Unsurprisingly, many women felt trapped and isolated within these suburban enclaves, despite the attempts to generate an air of neighbourliness through the provision of certain leisure facilities (Attfield 2007). Yet these neighbourhood units, primarily based around low-density single family housing, were intended to promote the virtues of a femininity in which a woman played the role of wife and mother, and was required to plan for the future in her capacity as a housewife. The modern house was thus the stuff of dreams:

> The woman in the home was encouraged to become a small-scale utopianist and a technician of space, whose work, just as much as in other more traditional kinds of housework, was to dream about and plan the perfect home . . . The modern home was construed as giving visual pleasure to its occupants, as well as allowing them to inhabit both inside and outside, amalgamating certain versions of the public and private based on a middle class identity. The housewife's task was to understand these latest ideas and adapt them to her family's needs in three important ways. First, women became increasingly responsible for tasteful home decoration, mediating the wild profusion of goods in the market to construe proper consumerism as an expression of 'self', a reflection of the well-cultivated inner person. Second, women's participation in the modern world was figured as the modernisation of domesticity, which entailed a fascination with the previously 'backstage' activities of cooking and cleaning. Third, the interior design discourse translated wider social changes into an everyday version of modernity particular to women's culture.
>
> (Lloyd and Johnson 2004, 270)

This modern inflection of a long-standing 'cult of domesticity', established in Victorian times, which encouraged women to seek satisfaction in fashioning their home, to be a loving wife and a good mother, cemented the imagined relationship between women and the domestic, private sphere.

The fact that many women embraced such ideals, and aspired to become domestic goddesses, can in many ways be related to the rise of a feminized consumer culture which was paraded in the retail spaces that targeted women: department stores, markets and, later, high street chain stores. Within such sites, items for decorating and embellishing the domestic interior were seductively displayed, becoming *fetishized* as objects which would facilitate the construction of the ideal home. Later, Ideal Home exhibitions, home improvement magazines and mail order catalogues impelled women to enter into a fantasy world of consumption. This increasing emphasis on the housewife as a consumer and the home as a site to be produced through rituals of homemaking is seen to have further alienated men from the spaces of the home, and increased their investment in the world outside. As such, the divide between the feminized domestic interior and a masculine exterior world was mirrored in men's retreat from the home in their leisure time, and their involvement in outdoor pursuits such as gardening. As Heynen (2005) notes,

Bird's-eye View of a Row of Patio Houses.
The individual gardens in front of the houses face a common green.

Figure 3.1 Housing as heteronormative? Artist's impression of post-war housing, Conder, 1949.

this was also reflected in the growing divide between the masculinized discipline of architecture, with its focus on exterior form, and practices of interior design regarded as soft and feminine.

These associations between femininity and domesticity gave rise to stereotyped notions that a woman's place is in the home. But the rise of new domestic appliances designed to be labour-saving produced a 'double-bind' for women: they raised expectations of cleanliness and impelled women to spend more time in the home (and often in the 'back spaces' of the home), further militating against their participation in the public worlds of work:

> Within the private spaces of the dwelling, material culture works against the needs of the employed woman as much as zoning does, because the home is a box to be filled with commodities. Appliances are usually single-purpose, and often inefficient, energy-consuming machines, lined up in a room where the domestic work is done in isolation from the rest of the family.
>
> (Hayden 1980, 172)

Many contemporary surveys of household reproduction show that the division of household tasks remains strongly gendered: women in nuclear households are much more likely than men to care for children, to cook, clean and do the laundry, whereas men are more likely to mow the lawn, deal with rubbish disposal and conduct repairs. Men, in most instances, take more responsibility for household finances and paying bills (Solomon *et al.* 2005). Developing this argument, Nast and Wilson (1994) argue that the networks of gas, electricity and telecommunications that crosscut domestic space register the importance of male labour and law, with the ordering of domestic space rendering women's work invisible. Production hence trumps reproduction, with the male labour valorized 'through wages' deemed to be more important than 'women's work' in the construction and the maintenance of the household (and the city) (Nast and Wilson 1994).

Research on household budgeting, homemaking and the division of household tasks thus begins to reveal the relationship between intimacy and economy that is negotiated within heterosexual coupled households (Duncombe and Marsden 1993; Illouz 1997; Zelizer 2007). However, there has been little work on the ways that the economic and emotional dependencies inherent in homemaking impinge on practices of sex itself. This omission is surprising given that most heterosexual couples setting up home do so, in part, to create a space for sex. Finding time and space for sex can be problematic for young people, especially in nations where there are traditions of filiation which mean children leave the home relatively late in their lives. Some may resort to commercial spaces set aside for 'socio-erotic' activities (see Case Study 3.1); others resort to surreptitious coupling in parked cars, on 'lover's lanes' or in public parks. But on obtaining a home, heterosexual couples acquire a space where they conduct intimacy on their own terms, free from the constraints present when they are flat-sharing or their parents are listening in. Morrison (2011) describes how couples develop a sense of belonging in their new home not just by removing the traces of previous owners, but through 'intimacies of space' in which sexualized contact and intimacy produces the home as a socio-erotic realm. This stresses that homemaking is the act of making domestic space 'homely' through the lived experience of that space as well as through material changes to it (Gorman-Murray 2007).

There is clearly much more that might be said about the way that the home is sexualized through practice, and about the way that the spaces of the home are eroticized through media representation (from the classical imagery of a couple making love in front of a roaring fire to pornography that depicts the kitchen as a sexual space). There is also much that needs to be said about the variety of sex that occurs within domestic spaces, noting that homes can be

CASE STUDY 3.1

Making space for sex outside the home: Japanese love hotels

Japanese 'love hotels' (*rabu hotero*) typically rent rooms not by the night, but for a *kyukei* (or 'rest' period) of one to three hours, making explicit the fact they are for couples to rent for the purposes of sex and intimacy. Growing out of the *tsurekomi ryokani* (tearooms) that served as legal brothels prior to the prohibition of sex work in the late 1950s, love hotels may be used for purposes of prostitution but serve a somewhat broader role as 'playful' spaces where couples, whether married or not, can experience a variety of forms of intimacy. As well as a bed and bath, the provision of TV with pornographic videos is standard, and most also provide condoms and vending machines selling sex toys and lingerie. The more expensive are veritable adult theme parks, offering rooms with feature beds, sunken baths, bondage equipment, entertainment centres and video games: some contain party rooms for groups, though few apparently tolerate same-sex couples. The entrance to the hotel is usually discrete and low-key, allowing customers to enter surreptitiously: anonymity is often ensured through automated lobbies and in-room credit card room billing: drive-in love hotels (known as motels) provide car parking spaces with curtains to obscure the residents' cars, lest these should be recognized (Basil 2008).

Estimates of the numbers of love hotels vary widely, but some suggest there are as many as 30,000 across Japan (West 2002), typically trading under names that give clues as to their nature: 'Joyful Club', 'Venus' or 'Hotel Lovely'. Some also boast elaborate and kitsch design, prominent within the streetscape as playful spaces. Legally, however, these were not distinct from other hotels until a 1985 edict placed stringent controls on their location by preventing them opening within 200 metres of schools, effectively classifying them as a sex-related land use (like sex shops and 'soapland' massage parlours). Despite this, 'extra-legal' love hotels continue to open in a variety of locations, often well beyond traditional inner city concentrations, simply by ignoring the 1985 definition of love hotels as providing rotating beds and wall mirrors of more than one metre square (West 2002).

Much rests on the location of love hotels in the city, with the playful escapism associated with these spaces produced via their placement in specific urban landscapes outside the everyday spaces of work and domesticity. Typically located in a peripheral space, like the temple compound of earlier times, for Chaplin the transition between the 'neutral' street spaces of the Japanese city and the neon-lit interior of the love hotel represents an important journey:

> The transition in the path from street to interior carries an additional load, since it is not simply a public–private threshold to be crossed, but one which entails a transgressive modal shift in societal terms: the couple entering are embarking upon a journey, a line of flight, which takes them from the prosaic, normalised context of their daily lives to something and somewhere else, and the prolongation of this passage from one to the other is a process of othering. The love hotel thus operates as a liminal environment, redesignating the street as a path into unknown territory, turning a mundane lunch-break into an experiential adventure.
>
> (Chaplin 2007, 21)

Moreover, for Chaplin the love hotel's interior design serves as something of a barometer of Japanese cultural beliefs, practices and preferences. As she catalogues, the love hotels of the 1960s were somewhat male-dominated spaces, their rooms aping Western images of a stereotyped bachelor-pad, complete with a Western-style raised bed, plush fabrics, feature headboard and wall mirroring, in stark opposition to the traditional Japanese sleeping arrangements where couples would sleep side by side – and not together – on futons. The quintessence of the bed as 'erotic stage' was reached in the 1970s as the rotating bed became standard in love hotels, this embrace of new technologies later manifest in vibrating and 'bodysonic' beds that responded to body movements. Subsequently, however, love hotels have become brighter and more open, partly responding to complaints they appeared somewhat seedy. Such is the acceptability of love hotels – with around one million couples visiting every day in Japan – that the design of hotel rooms has apparently raised and shaped aspirations regarding the appropriate setting for love-making in Japan, offering 'a whole visual, spatial and material vocabulary from which to draw' (Chaplin 2007, 65).

There are then multiple connections between domesticized sexuality and the spaces of the love hotel despite its apparent celebration of the illicit and immoral. As Basil (2008) notes, the love hotel was borne out of necessity, as most Japanese homes are small and have flimsy walls, with many couples living with children and elderly relatives in homes that lack private space. There is of course another sense in which love hotels help maintain family relationships in Japan: by providing a liminal space outside the home in which women and men can have illicit affairs, they also allow help maintain low divorce rates. For all this, Lin (2009) notes that they have become an increasingly important socio-erotic realm in which young women can construct self-identity by creating new domains of intimacy, loosening the connections between sexuality and marriage. Love hotels are hence contested spaces where public sexualities are negotiated, privately.

Further reading: Chaplin (2007); Cybriwsky (2011)

Figure 3.2 Love hotel, Shibuya, Tokyo (source: Enrique Dans/Creative Commons).

important spaces for the consumption of pornography (Juffer 1998), places of sexual experimentation and play (Storr 2002) and even group sex and 'swinging' (Worthington 2005). The fact there has been so little said about sex in the home arguably perpetuates the myth that heterosex is always monogamous, procreative and 'vanilla'. It also suggests that researchers remain squeamish about delving into people's private lives, assuming they will be reticent to share or disclose these experiences.

This academic squeamishness mirrors the state's insistence that it has no interest in the behaviour of consenting adults, and that there can be no public interest in preventing them from doing what they wish when out of public view. This was the justification for the decriminalization of male homosexuality in the UK in 1967 following the recommendations of the Wolfenden Committee, which suggested that it was not the state's business to interfere in questions of private morality, only to prevent behaviour which could be deemed as objectionable or obscene when conducted in public view (Mort 2000; Self 2003). In the much-cited words of the Committee, 'there must remain a realm of private morality and immorality which is, in brief and crude terms, not the law's business'. On this basis, their recommendation was 'that homosexual behaviours between consenting adults in private be no longer a criminal offence' and 'that questions relating to "consent" and "in private" be decided by the same criteria as apply in the case of heterosexual acts between adults' (cited in Weeks 1985, 67). Notably, the Committee also recommended a criminalization of street sex work that had the effect of encouraging the privatization of prostitution and an increase in home-based sex work (see Chapter Two).

The idea of the home as a zone of sexual privacy is one that is hence enshrined in legal as well as cultural understandings of the boundaries of public and private. In the US, the idea that the law should not interfere in the private sexual conduct of consenting adults was established in the 1960s in cases relating to the use of contraception within 'zones of privacy' (Hickey 2002). Though the Supreme Court's judgment in this case was about the use of birth control within a married relationship, this was quickly extended to unmarried couples on the basis that all individuals, whether married or not, had a right to privacy in the context of their sexual lives. Laws outlawing sodomy have also been repealed on the basis that if a consensual sex act is private, it is not a crime, though, interestingly, US courts ruled in 1986 that the rights to sexual expression and birth control only extended to heterosexually identified individuals. It was not until 2003 that a Supreme Court decision struck down all existing sodomy laws in the US; in Canada the Criminal Law Amendment Act repealed sodomy laws in 1969, but only to consenting adults over the age of eighteen where there were no more that two people present; in many other parts of the world, homosexual sodomy remains illegal even in private while heterosexual anal sex does not. Laws on consent for sodomy also often set higher age thresholds than vaginal or oral sex. The UK is exceptional in having set the age of consent for both heterosexuals and homosexuals as sixteen, and also making sodomy legal even if others are present (so long as these are not minors). Nonetheless, consensual bondage, domination and sadomasochism has often been subject to close legal scrutiny, with the state often intervening in the interests of promoting public morality and restricting harm: in the UK, the Spanner case of the 1980s triggered major discussion of the limits of the state's intrusion into private space (Bell 1995; Knopp 1997b). This is a discussion that continues in the context of people making and distributing 'extreme pornography' even when this is made with consent (see Wilkinson 2009; Attwood and Smith 2010; Johnson 2010).

Taken together, this suggests that the right to consensual sex is not always enough to guarantee a right to privacy, because courts can rule that certain acts, even if consensual, are still a crime. This means that in some jurisdictions, even sex toys and vibrators cannot technically be used: in Alabama, for example, 'auto-eroticism' and sexual stimulation is banned because of a court ruling describing it as detrimental to the health and morality of the state, which recognizes sexual acts only when these are, in their terms, related to marriage, procreation or family relationships (Herald 2004). Examples like this suggest that the right to sexual privacy is about morality and decency as much as it is about harm and consent. Putting it more simply, Hickey (2002) argues that a sex act can be considered as private only if it were one that a court would not object to seeing. While this makes many assumptions about the heterosexuality of law (Moran 2009) it reaffirms that legal definitions of privacy do not extend to encompass all sexual acts (Ashford 2010). Returning to a concept introduced in Chapter One, rights to sexual privacy are heteronormative in the sense that they are based on ideas of procreative sex as an inalienable right (Berlant 1997). Other transgressive acts remain subject to the surveillance of the state, being brought into public view and made 'live' through acts designed to ensure they are never truly private – even when homed.

The erotophobic landscapes of home: in what ways is the home sexually exclusive?

In this chapter, it has been shown that the idealized Western home combines particular ideas about love and intimacy with notions of commitment and a future based on the family. In the twentieth century, these notions of futurity mapped neatly onto dominant sexological and psychoanalytic discourses in which mothering and fathering were depicted as fulfilling social roles, with procreative sex deemed the most appropriate outlet for sexual energies (see Chapter One). Moreover, they supported capital accumulation by encouraging an aspirational form of homeownership from which mortgage lenders, real estate developers and the building industry benefitted. The nuclear family was thus integral to the industrial economy of production. In post-industrial cities, with the emergence of more varied living and working arrangements, pro-family groups have tended to stress the emotional rather than economic advantages of both marriage and homeownership, suggesting both are important in quality of life terms in an era when the state provides little safety net for those made unemployed. Contemporary society seldom hails the family or household as a productive unit, but rather sees it as a *welfare* unit providing emotional security, affect and a sense of destiny (Bhattacharyya 2002).

Heterosexuality's emphasis on futurity, inheritance, generation and coupledom has clearly made the house both an economic and *psychic* investment. Accordingly, this is an investment that people seek to protect, with the gating of many residential communities indicative of tendencies for homeowners to want to live in strongly ordered and exclusive residential spaces (see Aitken 1999). But even without gating, residential neighbourhoods can display strong exclusionary tendencies towards those perceived not to fit in, with Chapter Two having demonstrated that homeowners have been instrumental in Othering sex workers and seeking to exclude them from residential neighbourhoods. NIMBYism – the Not in My Backyard syndrome – is something that is noted to be particularly strong in areas where homeownership and family occupation is most marked, with possible incursions by sexual Others triggering strong exclusionary urges because of fears of moral, and sometimes physical, contamination. Wilton's (1998) study of community opposition to hostels for those living with HIV in Orange County reveals that homophobic discourse, combined with a general wish to avoid ill or diseased bodies, was prominent in the rhetoric of those local residents who sought to prevent the opening of such hostels (see also Takahashi 1997). Of all possible land-uses that can be proposed in residential neighbourhoods, hostels for those living with HIV feature as the least popular in many community surveys, some way below needle exchange centres, homeless hostels or hospitals of any kind (Dear 1992).

The idea that new facilities or developments might attract sexual Others is thus a key theme in oppositional rhetoric, even when those developments are not obviously sexualized. For example, an examination of public objections to a planned asylum centre in an English market town suggested that some residents felt the presence of unaccompanied male refugees in the towns would present a sexual risk for women and children, and that asylum seekers would be potential rapists (Hubbard 2005). This type of discourse is more overt where the premises are explicitly advertised as sexualized, such as might be the case for a sex shop, sex cinema or lap dance club. In such cases, homeowning residents in the vicinity can often make forceful claims about the possible impacts this will have in attracting undesirable elements, with dubious associations made between commercial sex and criminal acts such as drunkenness, drug taking, theft and rape (Edwards 2009). Interestingly, in many cases the leading figures in campaigns of opposition have been men who have claimed to be acting on behalf of women, invoking a feminized domesticated realm that needs protecting from the threat of predatory males (see Chapter Six).

The 'pervert' – the solitary and masturbatory male who is insufficiently socialized or domesticated – has been an important figure prompting many protests against the siting of adult businesses. In the contemporary era, this figure has often been regarded as synonymous with the paedophile, and those who oppose sex businesses emphasize the vulnerability of the children to would-be child molesters (Coulmont and Hubbard 2010). Moral panic about the paedophile, whipped up by a hysterical media, has made the protection of the child an unanswerable argument for the censorship and control of Other sexualities, and this is particularly the case when the sanctity and protection of the family home is endangered. A notable instance of this is provided by the resettlement or rehousing of registered sex offenders. In instances where such individuals have been exposed by the media, instances of vigilantism have followed as residents mobilize against what they perceive to be a threat to their neighbourhood (leading to many violent incidents, including a number of murders). Despite this, controversial legislation has allowed the publication and disclosure of the addresses of resettled and rehoused sex offenders: in the US, 'Megan's Law' was passed in 1996 following the rape and murder of a seven-year-old girl in New Jersey by a twice-convicted sex offender who lived on the same street, while in the UK, 'Sarah's Law' allows controlled access to the Sex Offenders register so parents with young children can find out if there are convicted sex offenders living in their neighbourhood. At the same time, more punitive laws have ensured that the reintegration of sex offenders in residential neighbourhoods is tightly controlled (see Case Study 3.2).

Fears of the sanctity of domestic space being threatened by sexual Others have also emerged in the numerous moral panics surrounding children's use of the Internet (Potter and Potter 2001). Contemporary social networking sites

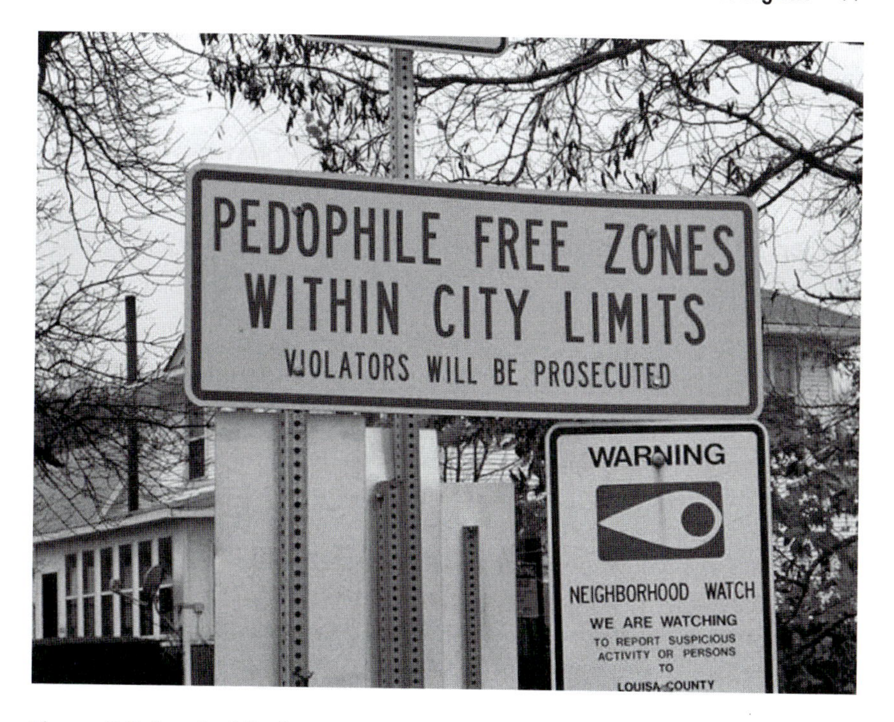

Figure 3.3 Paedophile-free zone sign, Louisa County, Iowa (source: Josh Smith/Creative Commons).

such as Bebo, myspace and Facebook are clearly popular with people of all ages, and have become a popular means for people to maintain social relations and intimacies at a distance (Valentine 2008). Yet there have been widespread concerns that such sites, which millions of young people are able to access in their homes and bedrooms via computers and mobile phones, provide a means for 'sexual predators' to insinuate themselves into the lifespaces of these young people. For example, the murder of seventeen-year-old Ashleigh Hall by a thirty-three-year-old man, who befriended her by posing as a nineteen-year-old, led to front page UK newspaper headlines posing the question 'Who's YOUR child talking to on Facebook tonight?' (*Daily Mail*, 10 March 2010), arguing that parents need to monitor their children's activities and react to them accordingly.

Here, the modern day 'folk devil' – the shadowy paedophile – is seen to be lurking on the Internet waiting to seduce the 'innocent' (gullible) victim, meaning that many recent campaigns for community protection focus on the need for parents to police the spaces of the home and ensure that they use appropriate Internet filtering so that their children are shielded from unwanted

CASE STUDY 3.2

Protecting 'family' spaces: sex offender restrictions in South Carolina

As well as community notification and electronic tagging, there are now many laws banning convicted sex offenders from living and working in areas in the vicinity of schools, day-care centres or playgrounds. With nearly three-quarters of a million registered sex offenders thought to be living in the US, and the 2006 Walsh Act demanding that authorities keep residency records for a minimum of fifteen years even for the least serious offences, and for life in most instances, the designation of spatial restriction zones is fraught with difficulty. Typically, restrictions forbid a convicted sex offender from living in the vicinity of a school or park, while some also forbid offenders visiting or loitering around swimming pools, sports centres, churches or bus stops (Grubesic *et al.* 2008; Wunneburger *et al.* 2008; Zandbergen and Hart 2009).

Spatial restrictions on sex offenders' residence in the US vary considerably, and though most stipulate around 1000 feet, Florida and California have laws that allow restrictions of up to 2500 feet from schools and parks. However, in some instances even wider spatial restriction zones have been implemented or mooted. For example, in South Carolina, a proposed Bill proposed a mile restriction zone, while a variation of it suggested a more conventional 1000-feet exclusion from schools and parks. Studying the potential impacts of these different restriction zones, Barnes *et al.* (2009) mapped nearly 7000 sex offenders' addresses in four metropolitan areas of South Carolina, using GIS (geographic information systems) methods to create buffer zones around child-centred facilities. Their analysis found that a one-mile restriction zone would have forced 80 per cent to move home and 92 per cent of vacant housing to be off-limits. Although the state ultimately adopted a 1000-feet buffer, Barnes *et al.* (2009) argue that even this restriction required 20 per cent of offenders to move, and made 50 per cent of the houses on the market in the cities inaccessible, raising major difficulties for accessing treatment and support networks.

Through the creation of large, overlapping restriction zones, sex offender laws can thus encourage the movement of convicted offenders away from more desirable, and socially organized areas towards more deprived areas; conversely, however, restriction laws can limit the affordability of housing options for offenders, and often make access to employment difficult (see Socia and Stamatel 2010). Spencer (2009) figures the sex offender as *homo sacer* – a notion invoked by Giorgio Agamben to describe someone who stands on the threshold between culture and nature, stripped of citizenship and reduced to a form of 'bare life'. The *homo sacer* is located within 'lawless' space, accused of being the opposite of all that is sacred, subject to sanctions that they do not have the means to challenge because they are deemed to have behaved in ways that deny them any entitlement to rights. Spencer (2009, 223) concludes that the sex offender is denied permanent residence because of the cumulative impact of spatial residence zoning, address disclosure and the threat of vigilante violence, and argues that 'the sovereign act of the expulsion of the sex offender from the community signifies the separation of a damned figure'.

Justified as a measure to increase public safety, the implementation of residential restriction laws has led some to surmise these are overly punitive and constitute 'retaliatory measures' (Socia and Stamatel 2010) and also fixate on fears of 'stranger danger' rather than the threat posed by sexual abuse within the family (Mulford *et al.* 2009). Moreover, some studies suggest that offenders generally travel to offend, and that proximity to children's facilities has not appeared to be a factor in the majority of instances of child sex abuse. This suggests that the geographies of sex offender residence in Western cities remain fiercely contested, and need to be considered in the context of attempts to define family space as sacred space.

Further reading: Barnes *et al.* (2009)

---

contacts. Such actions are demanded to prevent the seduction of young people, who are variously described in the media as 'vulnerable', 'naive', 'besotted', 'depressed' and 'lacking confidence' and quite unable to negotiate the spaces of the Internet (Potter and Potter 2001). Recent UK efforts to regulate 'extreme' pornography on the Internet suggest that while the paedophile's actions are seen as posing a danger to the nation, the blame for his existence must be shared by those who regulate and police the Internet. In many media accounts, paedophiles are depicted as not naturally evil, but corrupted by exposure to extreme images that are now easily circulated and stored in private spaces thanks to home computing. There is hence a common media discourse that talks of the 'failure' of regulation to protect society from the dangers of the Internet, encouraging parents to become involved in campaigns for heightened censorship of online sexual content, particularly when that content seems to escape the jurisdiction of national law (for example, see Wan 2009 on citizen actions in Hong Kong).

The moral panics surrounding cyberporn thus serve to reinforce particular notions of what it means to be a good parent in the contemporary era, and stress the need for vigilance to ensure that the home remains insulated from a sexual threat that is not 'out there' on the city's streets, but in the home. In all of this it is children who are depicted as innocent victims, ignoring the ways that minors might post descriptions or images of their own sexual activity, generate pseudo-images to create the idea that they are sexually active, or actively and creatively seek to negotiate sex with others via the Internet. Indeed, numerous instances of children striking up 'inappropriate' relationships via Internet chat rooms and role play environments like Second Life raise interesting questions about blame and culpability, and suggest there are very blurred boundaries between adult and childhood sexuality (Jewkes 2010). A phenomenon that has attracted considerable attention is 'sexting', which can involve people sending one another images of themselves in states of undress, performing sexual acts or making sexual suggestions, usually by mobile phone. This new application of technology came to public attention in 2008, when eighteen-year-old Jessica Logan committed suicide following months of taunting and bullying after nude images of herself that she had sent to her boyfriend were circulated, first across her Cincinnati high school and subsequently far beyond. In the US, a survey carried out by the National Campaign to Prevent Teen and Unplanned Pregnancy (2009) found that one in five teenagers had sent or posted online nude or semi-nude pictures of themselves and 39 per cent had sent or posted sexually suggestive messages. Research suggests sexting can create an uncomfortable combination of legal, social and emotional problems for participants, given they cannot be sure what receivers might do with the images once they get them (Chalfen 2009), and parents' response to this practice can often be extremely negative. The idea that children themselves are producing child pornography, and that laws forbidding the circulating of such images might be used to prosecute or punish young people suggests that the state and law needs to perform a series of intricate moves to distinguish between what is innocent, 'average' teenage behaviour and what is damaging and corrupting.

Though by no means a particularly urban phenomenon, 'sexting' has been woven into wider debates about the sexualization of youth in which sites of urban consumption are seen to be promoting an 'unnatural and premature' interest in sexuality (Coy 2009). Concerns here coalesce around the visibility of sexually explicit magazines, videos and clothes in retail environments (see Chapter Six), and the idea that these encourage girls and boys to value sex over intimacy. Examples include the marketing of *Playboy* product ranges to teen audiences; the easy availability of 'lads' mags' in high street newsagents and the use of sexual slogans in teenage advertising (Papadopoulos 2010). At the same time, TV programmes encourage young girls to dress as adult women, and vice versa (witness the BBC3 series *Hotter than my Daughter* which routinely dispenses advice on how too-frumpy daughters might compete with their more-sexy mothers). This has led some commentators to speak of 'corporate paedophilia', which Rush and La Nauze (2006, 7) describe as involving:

> the very direct sexualisation of children, where children themselves are presented in ways modeled on sexy adults. . . . The pressure on children to adopt sexualised appearance and behaviour at an early age is greatly increased by the combination of the direct sexualisation of children with the increasingly sexualised representations of teenagers and adults in advertising and popular culture.

It is of course possible to counter such blanket condemnation by suggesting that the problem here is not so much the visibility of sexual content and imagery, but a lack of content providing information about 'healthy' sexualities: certainly, much sexual imagery within mainstream pornography normalizes sexually subservient women and male sexual conquest (McKee *et al.* 2010).

In these debates, the home appears time and again as the one space where adults can shield children from inappropriate amateur or commercial sexual content, and introduce them to more 'acceptable' sexual knowledges – that is, those disseminated through school or the family (Hawkes and Egan 2008). The possibility that some pornography might offer more democratic knowledges and models of sexuality than official ones needs to be explored though: perhaps adult panics about young people's access to 'new' sexual knowledges are actually about an inversion of the normal power relations between adults and children where children are empowered and enlightened, and adults left in the dark.

So is the home an ordered sexual space from which inappropriate sex is easily expunged? Such debates would suggest not. Depictions of the home as a shelter from the sexual chaos of the world outside are too simplistic. The fact that the media persists in perpetuating the image of the ideal family home is significant, therefore, given this locates sexual threat in public space. This is an association that deserves critique given the levels of sexual violence reported within cohabiting relationships, as well as evidence that sexual abuse is more likely to be inflicted within the family than at the hands of a stranger (Pain 1991). For Jewkes (2010, 14), the 'powerful emotional and intellectual block' that prevents society recognizing that sexual abuse is more likely to occur in domestic space than elsewhere is evidence of the taboo status of incest, which is difficult for people to discuss in the context of domestic spaces imagined as loving and supportive. Turning the focus onto public rather than private safety hides the dangers of the domestic, and, as feminist scholarship has repeatedly highlighted, ignores the idea that the home is often a masculine domain where men seek to exercise sexual control of their family (and sometimes home-help and servants too) (Cox 2007).

'Other' homes and 'Other' sexualities: are all homes heteronormative?

Feminist critiques have been invaluable in showing homespaces to be restrictive and even dangerous environments for women in heterosexual relationships. Queer analyses push this further to argue that the spatial syntax and layout of the house embodies a *particular* set of heterosexual relations (Johnston and Longhurst 2009). For example, most housing assumes that family members will live, socialize and eat together in the same spaces, but that married couples will share bedrooms (and bathrooms) with their partner. Most houses are thus designed with three or four bedrooms: there is usually a larger, 'master' bedroom designed to accommodate a double bed and wardrobes. Increasingly, en suite bathrooms ensure that adult bodies can be cleansed, washed and dressed out of view of children. The normalization of these designs and layouts in national housing policy has helped institutionalize particular ideals of reproductive family living: what Warner (1991) terms *reprosexuality* (see Case Study 3.3).

For such reasons, much suburban housing has appeared unattractive for single people, as well as lesbian and gay couples. This has tended to encourage lesbian and gay residence in inner city areas, with classic studies of the formation of gay and lesbian 'villages' postulating that the incipient gentrification associated with gay and lesbian inner city residency resulted from childless single and partnered individuals seeking affordable apartments and flats (rather than houses) in inner city locations (Lauria and Knopp 1985; Adler and Brenner 1992; Bouthillette 1994; Hindle 1994; Knopp 1997a). As Doderer (2011, 434) argues, 'to have the chance to live in one's own apartment means to have a retreat of sorts, a shelter offering a distinct space from life outside . . . the chance to self-determine body, sexuality and identity'.

Here, the idea that inner cities are more diverse, cosmopolitan and accepting of different sexualities appears to have been a major factor underpinning the migration decisions of young single gay, lesbian and bisexually identified individuals. Conversely, the idea that small-town, rural or suburban living is alienating to lesbian and gay individuals has been reported in multiple studies

CASE STUDY 3.3

Normalizing reproduction: Singapore's family housing projects

Since its independence from the UK in 1963, Singapore's government has overseen a distinctive programme of citizen- and nation building in which the mantra of 'homes for people' has had considerable rhetorical appeal for the country's population. Oswin's 'queer' analysis of the activities of the Singaporean Housing Development Board seeks to expose the heteronormative nature of this housing, noting that 85 per cent of the population lives in apartment blocks designed and built by the HDB.

Given that dominant political discourse figures the family as the 'basic building block' of Singaporean society, Oswin suggests that the tenancy agreements underpinning the purchase of HDB flats are distinctly heteronormative, since any buyer has to be a Singaporean citizen, at least 21 and to have formed a 'family nucleus' consisting of the applicant and husband/wife, a divorced applicant with children or a married couple with children. Single parents who have not been divorced, gay couples and unmarried people are all prohibited from ownership. Given the virtual monopoly that the Singaporean state has over housing, this restriction of housing to 'family households' means that conversely, restriction laws can limit the affordability of housing for offenders, and foreign construction workers are only allowed to occupy dormitory accommodation.

Emphasizing 'the divide between heterosexuality and homosexuality is not all that goes into making [some]place heteronormative', Oswin (2010, 4) argues that Singapore's post-colonial aspiration to identify itself as a 'modern' nation-state institutionalized the idea of a 'proper' family nucleus. The 'happy' proper family was also deemed procreative – but not overly so – with the design and size of flats normalizing a small family rather than the traditional larger one. Interweaving particular ideas of biological reproduction, heterosexual comportment and nation building (Wong and Yeoh 2003), the Singaporean flat thus works materially and symbolically to normalize a particular form of reproductive heterosexuality, or reprosexuality (see also Lim 2004 on the marginalization of homosexuality in Singapore).

Further reading: Oswin (2010)

showing that feelings of discomfort in these spaces can often be traced to people's own experiences of 'growing up gay' in nuclear family households and living with the assumption of parent, siblings and others in and around the home that they are heterosexual (Brown *et al.* 2007; Gorman-Murray *et al.* 2008; Stella 2008). Most people are born into and grow up in heterosexual households, meaning that hiding lesbian or gay sexual identities from parents is commonplace for fear of disapproval, disavowal or even violence (Stella 2008). Even those who may have 'come out' to friends may seek to cloak this from family – something that Johnston and Valentine (1995) note may continue well into later life (with, for example, gays and lesbians 'de-sexing' their homes when parents visit so as to maintain a pretence of a heterosexual identity).

In Western societies there remains an expectation that there is a standard housing trajectory for young people. This involves moving from their childhood home to setting up a home of their own, possibly with a transitory period in between in which they live in shared accommodation. This trajectory is based on an idealized, ideological view of home which suggests that the shelter and support of family is gradually replaced by the support of housemates, and ultimately a partner, within the nurturing environment of the home (Mallett 2004). For someone growing up as lesbian, bisexual or gay, this transition can prove problematic, with even the temporary spaces of student halls of residence being heterosexualized in ways that normalize homophobic attitudes and language (Taulke-Johnson 2010). This implies that the majority of housing is scripted and regulated as heterosexual, and that it is difficult for gay and lesbian individuals to feel 'at home' in most contemporary housing:

> Gays and lesbians do not have a home to return to, either in historical terms or present day enclaves: most gays and lesbians live amidst their heterosexual families and neighbours without even the promise of a local community centre to find affirmation and support. Most grow up thinking they are the only ones in their communities. Even those who create homes are in a drastically different situation than that faced by racial, ethnic or national groups; such cultures and homes are created as adults, after the experience of isolation and rejection from one's own family and community.
>
> (Phelan 2001, 78)

So what happens when gay and lesbian identified individuals set up home? Research suggests there can be a strong imperative to reject dominant ideologies of the home, especially the asymmetric family implicit in the design and control of the nuclear family home. Gorman-Murray (2006, 148) suggests that this can include the use of homespaces in ways that resist heterosexual dominance, 'generating positive gay/lesbian identities'.

One dimension of this resistance to the heteronormative is that in coupled lesbian and gay households, there can be a progressive and egalitarian organization of domestic labour and finances, whereas heterosexual couples tend to divide tasks along traditional gendered lines. For example, in one US study (Solomon *et al.* 2005), married heterosexual women reported doing more of the household tasks than their partners did, including doing the dishes, cooking the evening meal, vacuuming the carpets, doing the laundry, cleaning the bathroom, doing the grocery shopping, ironing and taking the children to their activities and appointments. All of these tended to be shared by women in coupled lesbian households and men in gay households (cf. Kentlyn 2008). Though Duggan (2002) and others have suggested that civil partnership anaesthetizes queer communities and individuals into passively accepting forms of inequality in return for domestic privacy and the freedom to consume, Solomon *et al.* (2005) found few differences in housework sharing in situations where lesbian/gay couples were in civil unions and those where they were not. Gorman-Murray (2006) notes other important differences between cohabiting lesbian/gay and heterosexual households, with a tendency for the maintenance of separate bedrooms in gay/lesbian households. Despite this, his study reported that shopping for furniture and decoration of shared spaces was considered vital for consolidating gay and lesbian coupled relationships.

However, the literatures on lesbian, gay and bisexual negotiations of home hint at some important differences between lesbian and gay practices of homemaking. Some of these seem based on possibly flawed assumptions that men are more spatially expansive and outgoing in their lifestyles, and that women are more domo-centric (home-based). In essence, a common argument is that lesbian social networks are much less based on commercial spaces and spaces of night-life (pubs, clubs, restaurants), being constructed around 'homed' networks. While lesbian neighbourhoods and 'communities' are sometimes visibilized, this tends to create communities that are much less overt than the gay male villages most noted in studies of sexuality and space (see Chapter One). Nevertheless, for many lesbian identified women, home is also a site where they make their sexual identity visible in a conscious attempt to challenge assumptions of heterosexuality and to contest societal pressures to hide their sexuality:

> These homes are places in which to nurture, maintain and actively assert lesbian identity. At the same time, the lesbian home encompasses contradictory meanings. The home might be a place of affirmation, a place where a lesbian feels most comfortable expressing her sexual identity.

> Simultaneously, it is often a place to which such expressions are rigidly confined by societal disapproval and harassment. Some of the same contradictions exist in lesbian neighborhoods. While these areas might be places where lesbians form secure, affirming communities, they might also be places where lesbians are oppressed.
>
> (Elwood 2000, 12)

However, the notion of the lesbian neighbourhood makes little sense in most small cities and towns, where lesbian women tend to live in more conventional residential 'family' and suburban spaces (Valentine 1996). This contrasts somewhat with some of the descriptions of gay men's housing choices, with multiple studies focusing on the pivotal role of gay men in processes of inner city gentrification (Knopp 1997a). This suggests a particular set of housing preferences among gay men and cohabiting gay male households, with a privileging of inner city lofts, flats and apartments which reflect both a set of aesthetic preferences as well as demand for a location that allows easy access to city centre sites of consumption and night-life. The idea that gay men are significantly involved in the renovation and re-aestheticization of inner city areas contributes to popular stereotypes that gay men are domestically stylish (Kirby and Hay 1997; Leslie and Reimer 2003), something highlighted in television shows such as *Queer Eye for the Straight Guy*, where gay men offer advice to heterosexual men on how to makeover their lives and homes (Hart 2004).

As such, despite the frequent representation of gay leisure space as the key site for constituting non-heterosexual identities, it is clear that domestic spaces can be massively important for gay men too, and theoretically allow for the development of alternative or queer masculinities out of the gaze of heteronormative society. This is often reflected in the ways that gay men display their aesthetic sensibilities (see Gorman-Murray 2007; Gorman-Murray *et al.* 2008 for vignettes of the importance of displaying 'gay possessions'). This 'queering' of domestic space both eradicates any trace of former heterosexual owners of the home, and also distances gay men from the inherent heteronormativity of domestic space (Gorman-Murray 2007). Notably, the styles stereotypically associated with gay gentrification are not those associated with homemaking as performed by heterosexual women, and reject feminized and soft styles in favour of a modern aesthetic that is harder lined. What is notable here is that the stereotyped aesthetic and stylistic choices of gay men are being adopted by heterosexual identified men, thanks in part to design magazines like *Wallpaper* which appeal to the young male urban professional. This emphasis on displaying a discerning taste, and an associated interest in fashion, food and grooming is arguably one that took root in metropolitan centres where gay gentrification was most pronounced, and led to the identification of 'metrosexual' lifestyles. This normalization of gay aesthetic styles and tastes, and their adoption not just by the bi-curious or queer, but by 'straight' men, suggests that there has been an assimilation of gay male lifestyles – something that supports the idea that gay cultures have been eviscerated through a selective incorporation in the mainstream (an argument explored further in the Conclusion).

Consideration of gay and lesbian households sheds fresh light on homemaking practices, and reveals the inequities that adhere to many heterosexual households. But there still remains much that might be said about the sexual lives played out in other types of household. For example, little has been said in the literature on sexuality and space about single households, despite the increasing rates of singleness evident in post-industrial societies. Those who live alone are, according to Budgeon (2008, 309) often seen as 'selfish, deviant, immature, irresponsible, lonely, unfulfilled, emotionally challenged, lacking interpersonal ties and strong social bonds', with 'very few positive representations of uncoupled lifestyles available as resources for people to draw upon in constructing a positive identity'. Cobb (2007) concurs, suggesting that the 'coupled' society imagines single people living alone as desperate and lonely, with single housing being a transitory form one inhabits only after one coupling finishes and before another one starts, and seldom out of choice. But while Cobb notes the widespread nature of such rhetoric, with media representations creating new anxieties about singleness, it could be argued that new representations of the man (and woman) about town have generated some counter representations of single housing. Lloyd (2008) demonstrates this in her study of the marketing of inner city housing in Sydney to single, career women. Herein, she focuses on the advertising of the Lumina Apartments, inner Sydney, which sold a sexy Sydney lifestyle in which single women were depicted as using the space of the home not as a space for housework, but for consumption – and especially sexual consumption (one advert coyly represented a woman masturbating while watching TV, home alone). Using appropriately sexy models, these adverts hinted at a *Sex and the City* lifestyle, knowingly implying that this is an inversion of 'normal' gendered relations, but also, Lloyd (2008) argues, easily within reach. In her studies of gentrification in Canadian cities, Kern (2007) likewise notes that condominiums have been marketed as offering access to the social and economic advantages of homeownership from which single women have often been excluded, with inner city dwelling allowing them to combine work and leisure in an environment rich with urban spectacle (and sex). While she argues that this does not fundamentally challenge heteronormative assumptions about what women want from housing, this certainly chimes with Leslie and Reimer's (2003) observations that urban lifestyles are being sold not just to couples but to single mobile young professionals – including women.

Hence, while the 'bachelor' or man about town provides an interesting case study of domestic masculinity (Osgerby 2005), the 'bachelorette pad' has become equally privileged as an acceptable and stylish alternative to the nuclear household. Embodying an 'ideal' of stylistic expression and commodity consumption, modern flats designed for single people are decorated, accessorized and mediated as socio-erotic spaces, suggesting that the interior spaces of these private realms can be important in invoking particular associations between consumption, sexual performance, fertility or virility (Fraterrigo 2008). The domestic realm can hence be sold as a space of *seduction*.

The demographic and sexual shifts of the late twentieth and early twenty-first centuries have accordingly ushered in new understandings of the sexualization of homespaces, opening up an increased range of housing options for those pursuing sexualities that lie outside the framing of the idealized nuclear family. Despite this, and the decline of marriage in most of the urban West, it is still assumed that people in long-term relationships will instantiate their coupledom by moving in together: as Roseneil (2006) argues, if they do not do so, it is assumed they are not partners at all. The estimated four million men and women who 'live apart together' in the UK (Haskey 2005) thus encounter considerable social pressure to cohabit, while those seeking to develop non-traditional polygamous or polyamorous household formations find that dominant housing forms are rarely adaptable to their needs. Rubin (2001), for instance, suggests that while there were experiments in new housing forms in the 1970s based on notions of comarital relations and 'swinging', polyamory remains difficult to enact. Cities continue to offer housing predominantly marketed at families, couples and individuals, which is far from ideal for communal living, and queer experiments in developing other ways of living in squats and communes have often floundered in the face of gentrification or the drift of participants towards 'compulsory monogamy' (Conover 1975; Overell 2009). Tellingly, many utopian non-monogamous communes have been established in the rural, with participants often justifying their choice to live in a non-normal household as part of an attempt to create both emotionally and environmentally sustainable communities (Klein and Anderlini-D'Onofrio 2005). This suggests that the city encourages neither the making of sustainable housing nor households that allow for combinations of multiple partnering, non-monogamy, multiple intimate friendships or the celebration of 'free love' (see also

Valentine 1997; Kirkey and Forsyth 2001 on lesbian and gay separatist communities).

Conclusion

While the domestic can extend beyond the home to encompass the neighbourhood, the town, the city and even the nation (Marston 2004), this chapter has focused on the household itself as a site where particular ideas of sexuality are normalized. In Western societies, one of the characteristics of the household has been its assumed heterosexuality; conversely, one of the defining norms of heterosexuality is that it 'involves long-term monogamous relationships in which partners share living space' (Van Every 1996, 41). Yet this chapter has suggested that this equation between heterosexuality, coupledom and the home is becoming more tenuous in an era when housing is explicitly marketed to single young people, including lesbian or bi individuals. But a consideration of the unsuitability of most housing for those rejecting coupledom, seeking to live in communal arrangements or having a polyamorous lifestyle underlines that at the heart of many households is a couple.

So does this emphasis on the couple, rather than the heterosexual couple per se, suggest that the Western city has begun to accommodate a broader range of sexualities? Perhaps. Yet this conclusion needs to be tempered by the observation that this is not so much a celebration of sexual Otherness, rather a normative recuperation and commodification that has domesticated homosex, and sought to push it squarely into the private realm. Through its association with domesticity, coupling and parenting, homosexuality can be channelled to 'socially beneficial' ends, something reflected in forms of housing designed to enable and encourage the sharing of lives between individuals drawn together by a desire for intimacy rather the desire for sex itself. This suggests a complex entwining of sex, intimacy and subjectivity in the spaces of the home, as Oswin and Olund note:

> The nonsexual aspects of intimacy predate sexuality as the truth of the modern self, and they persist into the present. The heterogeneous elements that proper intimacy and sexuality currently comprise are certainly overlapping sets – the body, the self, ideally the home, to name just three – but they are not coterminous . . . As the range of acceptable disciplined sexualities increases (even if their degrees of legitimation differ, e.g. civil partnerships versus marriage) intimacy per se is coming to the fore as a regulatory construct.
>
> (Oswin and Olund 2010, 64)

This increasing commodification and privatization of intimacy (which Valentine 2006 suggests is also virtualized) has implications for individuals' sexual safety given the home is not necessarily a realm where sex is practised safely and with consent. As this chapter has demonstrated, while the actions of homeowners are often intended to repel those seen as sexually dangerous (most notably, the paedophile) the home itself can be a space of rape, domestic violence and incest. This given, it appears that homophobic/erotophobic arguments (that sexual dissidents should remain invisible) and assimilationist arguments (that they should conform to mainstream values) combine to reproduce the divide between public and private, with transgressions of this divide retaining the power to shock. This theme of transgression is one that will be highlighted in the next chapter, where the focus shifts from sex in private to sex in public.

Further reading

Aitken (2009) offers an alternative take on dominant notions of parenting, care and housework as 'women's work' in his exploration of the 'awkward' spaces of fathering.

Blunt and Downing (2006) is an excellent introduction to literatures on the geographies of home, with particular attention paid to feminist perspectives.

Gorman-Murray (2006) provides a useful case study of homemaking among both gay and lesbian communities, drawing on narratives collected from respondents in Australia.

4 Public sex

Learning objectives

- **To understand the importance of the public–private binary in structuring ideas about sex and sexuality.**
- **To appreciate the transgressive potential of sex when it occurs 'out of place'.**
- **To recognize how (legally defined) distinctions of private and public blur in the realms of sexual practice.**

In the last chapter, the assumption that people can perform their sexualities freely in their home was questioned. It was shown that, contrary to expectation, the state's interests in sexuality often extend into the private bedroom, making 'private morality' a matter of 'public concern'. Yet when the domestic realm does offer some respite from the intrusive gaze of the state and law, it was noted it can become a space where sexual violence and coercion can be practised with some degree of impunity. Homespaces are thus highly ambivalent spaces of sexual belonging, and while they can offer some shelter from dominant, heteronormative values, their emanicipatory potential remains doubtful.

This chapter moves to a set of spaces that are the seeming obverse of private spaces, being denoted as 'public'. Classically identified as those spaces that exist outside and between the realm of work and home, these include the streets, squares, pavements, car parks, beaches, green-spaces and wastelands which are knitted into the urban fabric in a variety of ways, whether loosely or otherwise (see Franck and Stevens 2007). Providing sites for life 'between buildings', they are often seen as vital by urbanists as they provide spaces where difference cannot be avoided, and must be negotiated. Richard Sennett (1970), for example, famously wrote of the 'uses of disorder', and the importance of public spaces that, through tendencies towards anarchism, force people to deal with one another. Since Sennett's pivotal intervention, opposing the *purification* of public space has been a key trope in urban writing, with the privatization of many public spaces enacted by CCTV and ever-present surveillance accused of negating the possibilities for free association, open discussion and spontaneous play (Mitchell 1995).

But questions remain about how much disorder is actually desirable. As Iris Marion Young (1990, 240) noted in her work on the cultural politics of difference, when entering public spaces one 'risks encounter with those who are different, those who identify with different groups and have different forms of life'. While senses of neighbourhood, community and urban conviviality are seen to develop out of such encounters, the streets can of course become too lively, with the anarchism that Sennett espouses threatening to spill over into conflict and violence. Even those arguing that 'struggle and strife' are as important as 'joy and enjoyment' in authentic urban public spaces (e.g. Merrifield 1996) concede that there need to be some rules shaping the terms on which public space is occupied and used, noting that much urban life can be quite 'grisly'. There are, as Merrifield notes, certain dangers in romanticizing the grittier elements of street life, or suggesting that 'anything goes' in public space. Others, like Sennett, dissent from this. For Marxist urbanist Marshall Berman (1986, 482), for example, 'the glory of modern public space is that it can pull together all the different sorts of people who are there . . . [*and*] can both compel and empower all these people to see each other, not through a glass darkly, but face to face'. Berman's (1986: 484) essay on street-life suggests that public space fulfils a fundamental role in the life of cities by facilitating the meeting and mis-meeting of people of varied 'classes, races, ages, religions, ideologies, cultures, and stances towards life', and argues that encounters with those of other sexual dispositions are part of this. Holding out great hope for the development of a more liberal and authentic sexuality forged in the spaces of the public city, Berman's essays on New York street-life often lament the 'cleansing' of the streets under Mayor Giuliani, and regret the disappearance of the street hustlers, porn peddlers and prostitutes once characteristic of areas like Times Square (Berman 2001). For others, this ignores the deep discomfort some experience when confronted with the sights and sounds of sex (Listerborn 2004).

Poised between order and disorder, the public spaces of the city are therefore never quite as liberatory as might be supposed:

> The street is the site of confrontation between different social groups in the city, and between those groups on the one hand and the forces of law and order on the other. From the point of view of the custodians of the law there is always *trop de vie*, too much life on the streets. They therefore

> develop mechanisms which are supposed, above all, to restrict the manifold functions of the street, strategies producing clarity and order-creating spatial division.
>
> (Schlör 1998, 33)

Legal attempts to exclude or curtail some activities from public space on the grounds they should be limited to the private realm means that some spaces which appear public are actually designated legally as private. This suggests that a *topological* understanding of public and private space (i.e. that privacy can always be found in private and publicness found in public) is flawed: instead we need to consider the ways spaces *become* public or private through acts of ordering and law making (Iveson 2007).

This chapter therefore focuses on the ways that the state and law seek to push sexual acts and practices 'out of sight', and into private. In doing so, it outlines the processes that have effectively served to de-sexualize public space. Yet at the same time it notes that there remain many sites outside home or work where people can still claim a transgressive, alternative or 'queer' sexual identity. Exploring the political and cultural significance of such sites, and the sexual exclusions and inclusions associated with them, the chapter highlights the ambivalence of public transgression. Noting a variety of related attempts to take sexual politics to the streets, this chapter will emphasize some of the problems of radical public sex, describing how it is policed through laws relating to public decency and obscenity. The chapter begins, however, by asking why sex in public has come to be seen as impolite and inappropriate.

Impolite bodies: in what ways is public decency related to sexuality?

Questions of what we can do with, and to, our bodies have always been central to attempts to define and discipline sexuality. Foucault's work, for example, showed that the 'anatomo-politics' of conduct has been as vital as the biopolitics of population management in the regulation of sexuality (see Philo 2005). Crucial in this process have been state programmes of education, welfare and social policy designed to promote clean, healthy and (re)productive bodies. Sex education is one part of this process, imparting knowledge that entwines social, economic and moral assumptions about the gendered body, ultimately defining what is considered sexually normal and healthy (Mort 2000). Such programmes of sex education have been institutionalized in 'places of formation' – schools, libraries, doctor's surgeries and family planning clinics – where people are encouraged to exercise forms of self-surveillance and bodily management which render them legible as disciplined sexual subjects. Historically, for those 'dangerous' individuals who refuse to conform to these sexual norms, 'places of reformation' such as hospitals, asylums and Magdalene institutions awaited, ready to 'treat' and repair the recalcitrant, encouraging them to conform to norms of sexual hygiene and order (see Smith 2004, for example, on homes for unmarried mothers in mid-twentieth-century Ireland).

Though highly distinctive in many respects, Foucauldian analyses of the sexual body share much common ground with the figurational analyses inspired by the work of Norbert Elias on the social and cultural meanings of the body (Smith 1999). Charting a history of manners and styles of bodily comportment, the historical analysis provided by Elias emphasized that understandings of bodily management have varied over time and space. He suggested that a key trait of modernization was the idea that people needed to distance themselves from primitive societies by changing the ways they used their bodies and adopting civilized ways of acting towards other people (as well as to animals and nature). He suggested that contemporary notions of what is taken to be a proper and 'normal' body can only be understood by examining the 'civilizing process' that radically changed understandings of how human beings were supposed to act. Crucial here was the Enlightenment, an era associated with philosophical and scientific developments that challenged long-standing ideas that people's place in the world was 'god-given'. Arguing that reason was the basis of progress, and that humans were distinguished from the rest of the animal kingdom by virtue of their capacity for improvement, Enlightenment thinking ushered in new understandings of civility and manners. As Elias' work on civilizing processes describes, this saw post-medieval European standards regarding violence, sexual behaviour, bodily functions, table manners and forms of speech being gradually transformed by increasing thresholds of shame, repugnance and disgust. For example, he noted that in medieval times people would blow their nose with their fingers, adopting the 'civilized' handkerchief only in the seventeenth century. Similarly, he reports that it was common for people to talk to each other while defecating or urinating in public communal areas (such as fields and streets) until late medieval times; and to eat with their hands until the seventeenth century (see Jervis 1999).

Prohibitions concerning 'sexual display' were a crucial part of this process. In spite of official Christian doctrine, most people seemed relaxed about public nudity in the Middle Ages, but the gradual segregation of sexuality from social life meant that this type of behaviour was viewed as increasingly embarrassing, with children seen as in particular need of protection from the 'shameful' sights and sounds of sex: Elias (2000, 148) contends that by the twentieth century there was an almost complete conspiracy of silence surrounding 'the sexual area of the life of drives' when in the company of children. Elias (2000, 134) thus concluded that 'the feeling of shame surrounding human sexual relationships' became 'noticeably stronger in the civilizing process' over time. As Jervis (1999) writes, sex is now surrounded by complex proscriptions and exclusions because it reminds us of our irreducible baseness, and our 'uncivilized roots' in nature, meaning that it must be carefully controlled so as to distance ourselves from nature (see also Brown and Rasmussen 2009 on the prohibition of bestiality).

In *The Civilising Process*, Elias (1978) thus outlined a theory of the body that explored how understandings of civility transformed over the centuries to the point where 'modern' notions of bodily comportment and 'good taste' reigned supreme, and where to ignore these notions was to appear uncivilized, and little better than the animal. In effect, the body became (re)imagined as something that needed to be managed by its 'owner', moulded so as to conform to ideals of middle class comportment and politeness. In this sense, poorly managed bodies had an increasing capacity to provoke disgust, particularly where the 'owner' of that body seemed unwilling or unable to effectively manage its boundaries, and maintain a clear distinction between inside and outside. In contemporary, modern societies, bodily fluids have a particularly pronounced capacity to generate feelings of disgust: saliva, shit, sperm, blood and urine are taboo. The fact that many of these substances are exchanged in sex acts means that sex itself has become depicted as a potential realm of risk in which one person is endangered through contact with the other via sexual fluids. Moreover, as Mary Douglas' (2002) classic study *Purity and Danger* suggested, such risks tend to be regarded as asymmetrical, with men often seen to be threatened by women, as 'leakiness' has been an enduring signifier in representations and understandings of female bodies (see also Grosz 1994). Such pollution taboos have resulted in sex acts being surrounded by complex cultural conventions about when and where sex can happen. In short, like defecation, menstruation, urination and other acts that compromise the boundaries of the body, sex itself is supposed to occur only out of public view:

> Erotic reality is deliberately set off in time (evenings, nights, weekends, vacations); it is segregated in space (home, bedroom, cat-house, secluded beach and back seat of a car); it is stage set (drawn curtains, dimmed lights, burning logs in fire place, mood music, sexy clothes); it is aided by the intake of special foods, alluring scents (perfume, incense) and mood altering substances (liquor, certain drugs).
>
> (Weitman 1999, 75)

This means that there are strong taboos in Western societies about sex that occurs in the wrong times and spaces – taboos that can appear magnified in other cultural contexts (see Case Study 4.1). These taboos are often mirrored in legislation that defines public sex as an offence against public decency: in the UK, for example, the Sexual Offences Act (2003) does not legislate specifically against sex in public space, and allows sex in places where the participants might have a 'reasonable expectation' of privacy. On the other hand, if sex occurs in a 'private' space (such as a garden) but is witnessed by passers-by, this would be deemed an outrage of public decency. This is irrespective of whether the viewer was in any way corrupted or depraved by the conduct, meaning that it constitutes indecency without necessarily falling into definitions of obscenity (see Chapter Six).

But codes of manners are not simply about the spaces where sex itself can be made visible, containing an assortment of ideas about how the relationship between sex and intimacy should be expressed bodily. Wouters (1987) follows Elias in exploring the 'modern manners' books and media that have constructed the idea that it is deeply unseemly to seek sexual gratification without first cultivating an enduring intimacy. The early twentieth-century manners books that Wouters explores map out a beguiling set of guidelines relating to appropriate bodily comportment when courting, suggesting appropriate ways of dressing, dancing, kissing and 'petting'. While such manners books have arguably subsided in their significance, and society has become more informal, Wouters suggests that there is an increasing emphasis on emotional management and self-restraint in Western societies, as witnessed in the number of 'self-help' books that encourage people to relate to their sexuality in particular ways (see Albury 2002 on the rise of the 'sexpert'). The upshot is that people are encouraged not just to make their bodies legible within what Butler (1993) terms the *heterosexual matrix*, but to perform their sexualities in ways that acknowledge the cultural privileging of romantic love over sexual lust, chastity over promiscuity and bodily restraint over excess.

In other considerations of the sexed body, psychoanalytical theories about the importance of preserving self-identity abound, with *abjection* being a widely deployed concept. As outlined in French feminist psychoanalyst Julia Kristeva's (1982) influential *Powers of Horror*, abjection is the process in which the boundary between Self and Other becomes blurred, and in which the individual seeks to cast off or repel that which disturbs the distinction between his/her body's inside and the world outside. The abject body is one that leaks wastes and fluids, violates its own borders and does not conform to social standards of cleanliness or propriety. Disgust is the primary embodied reaction to encountering the abject body, with geographer David Sibley

(1995) consequently arguing that the urge to exclude the abject from one's proximity is perfectly explicable given our desire to maintain bodily cleanliness and purity. Bad objects, and bad bodies, are thus distanced through processes designed to purify or sanitize: as was explored in Chapter Two, historically prostitutes have often been cast out to marginal spaces because of their alleged diseased and dirty bodies. Such theories are also relevant to explaining the social and cultural antipathy towards the undressed body, which is regarded not necessarily as dirty but as potentially defiling and disgusting because of the exposure of parts of the body coded as sexual. Hence, if an individual is seen to cast off its everyday self and shift into an 'erotic embodiment' (defined through exposure of the erotically charged parts of the body, such as the anus, breasts, penis and vagina), this can provoke feelings of revulsion and disgust (see Hubbard 2000).

These different theoretical perspectives all lead to the same conclusion: the undressed body appears undisciplined, uncivilized or disgusting when encountered in public. However, this is complicated by the idea that the naked body may be acceptable in public view if it is presented as an artistic statement (in dance or theatre, for example). Yet in such cases, the naked body is generally represented not as a grotesque or disgusting body, but as a classical, self-contained, cleansed body – the *nude* as opposed to the naked (Nead 1992). In other cases, as Barcan (2004) relates, the dialectic of clothed and naked maps onto established understandings of what is social and what is profoundly anti-social. Nudity changes 'the familiar boundary between body and world, as well as the effects of the actual gaze of others and/or the internalized gaze of an imagined Other' (Barcan 2004, 24). Moreover, given the persistence of a strong set of discursive connections between nudity, sex and immorality, the naked or unclothed body is often regarded with suspicion, and legislated against as breach of accepted standards of decency (Mason 2005).

However, decency laws are far from consistently applied, with Barcan (2004, 110) arguing that 'the nakedness of real-life male bodies in public space is much more "dangerous" than that of female bodies'. Put simply, this is to argue that when women are naked in public view they might be considered as 'morally dangerous' but are less likely than men to be viewed as criminal or deviant. These gendered assumptions mean that public male exposure is far more likely to trigger a legal response, with the histories of prosecutions for indecent exposure suggesting that the state and law has a particular interest in protecting 'innocent' women and children from the sight of male nudity. An example of this is provided by some of the debates around bodily comportment on city beaches. Located at the edge of cities, between land and water, beaches are classic *liminal* spaces, poised between nature and culture

CASE STUDY 4.1

Causing public offence: sex on the beach in Dubai

Dubai (United Arab Emirates) is one of the 'shock-cities' of the twenty-first century, its hyperurbanization based upon a combination of oil-based wealth, tourism and intense property speculation. Eighty-five per cent of residents are migrant workers, with a large majority of these low-paid male labourers and taxi drivers from the Indian subcontinent. While its neighbouring state, Saudi Arabia, has ostensibly harsh laws concerning displays of public affection, nudity and homosexuality, policed by the Saudi Mutaween, the UAE's reputation for homophobia seems based around the media's frequent condemnation of same-sex marriage and parties. While homosexuality is illegal in UAE, at 'street level' the reality can appear somewhat different, with the multicultural ethos in Dubai encouraging forms of homosocial and homosexual encounter among the male-dominated workforce. Ingram (2007) suggests that many of the local Emirati also engage in same-sex practices, with key public spaces – most notably the freely accessible Jumeirah Open Beach – being important sites in the making of regional and local gay support networks in a country where many Internet sites relating to gay or queer cultures are blocked. Largely out of the gaze of women, and at dusk and dawn in particular, Open Beach serves as a surreptitious space of male sexual encounter, from where men will retreat to hotels, backstreets and cars to have sex (Ingram 2007).

The fact that homosexuality is evident in Dubai is interesting given the attempts of the ruling Emirati to promote the city as multicultural, tolerant and welcoming. Though male dominated (70 per cent of the population are male), the city is keen to appear a welcoming destination for Western women. As Smith argues, this relies on projecting images of *safety*:

> Dubai bills itself – against the perception of the region . . . as one of the safest cities on Earth and as a haven for personal freedom. It is a place where, allegedly, you (no matter who you are) have nothing to fear. This is a city of 1.3 million persons where those with even small amounts of money – unlike Saudi Arabia – are legally free to dress as they please. This is because, in trying to meet its ambition to become a globally significant city, Dubai has come to rely on foreign labor – not just in construction and domestic service, but in management, retail, IT, engineering, medicine, and practically any other sector one could imagine.
>
> (Smith 2010, 270)

Smith continues by noting that instead of forcing non-national women to wear a veil to protect them and keep men 'un-aroused', as is the custom in the region, Dubai's authorities have instead decided to police the behaviour of men. To these ends, the Al-Ameen Service has been set up, an English language helpline that is advertised as being primarily for women to call if they are being stalked or harassed. Those guilty of sexual harassment are 'outed' as sex pests in the national media through a form of naming and shaming supposed to deter others. Hence, police regularly patrol mixed beach spaces, arresting men for 'ogling' or photographing women bathers. In the words of Colonel Al Za'abi, the 'aim is to make beachgoers feel comfortable and protected from groups of repulsive men as tourists visiting Dubai's beaches grow' (cited in the *Gulf News* 2009).

As such, the authorities in the city continue to make efforts to make the beaches more 'family' friendly and designate women-only days on Jumeirah beach. It is in this context of the desire to be seen as a tolerant, open and safe space – especially by women – that the arrest of two expatriate British workers for having 'sex on the beach' in 2008 caught the attention of the world's media. Some hours after having been at a champagne brunch for expatriate workers at Le Meridien hotel, Michelle Palmer, 36, and Vince Acors, 34, went to Jumeirah beach by taxi, where, it was later reported by police, they were warned to stop their amorous behaviour. When the police returned some time later they were reportedly having sex on the beach. Charged with having unmarried sex, committing public indecency and insulting a police officer, the pair were sentenced to three months in prison and fined 1000 dirhams (at the time, approximately £160 or $350).

This story was widely portrayed in the Western media as encapsulating a culture clash between Western permissiveness and Islamic repression, with the seeming transgression of appropriate sexual comportment by the pair taken as evidence of a more fundamental division between debauched expatriate lifestyles and ascetic Emirati cultures. The fact that the couple could have been arrested in most Western nations for similar public behaviours was rarely noted in such reporting, suggesting the perpetuation of an overly simplistic binary between Western and non-Western sexual attitudes. Nevertheless, the intense interest in Michelle Palmer and Vince Acors suggests that the sexualized body, when encountered out of place, can prompt considerable anxiety.

Further reading: Smith (2010)

(Shields 1991). Despite the efforts of the authorities, city beaches have often become 'carnivalesque leisure spaces of ritual inversion of the dominant, authorised cultures' (Daley 2005, 155), with public bathing having presented something of a conundrum for the authorities. On the one hand, medical practitioners have suggested bathing has major therapeutic benefits, and improves the health and well-being of urban citizens. On the other, understandings of appropriate comportment discourage the display of the undressed body in public. Historically, even the sight of the 'scantily clad bather' provoked anxiety (Booth 1997): throughout the urban West, beaches became subject to regulations that forbade 'sunbathing' except in designated areas, with changing booths and segregated areas ensuring standards of modesty and propriety were maintained. In the case of Sydney's city beaches in the mid nineteenth century, White (2007) describes 'respectable' politicians complaining vocally about the ungentlemanly behaviour of working class 'larrikins' who 'sexualized' the beaches through their overt and 'exhibitionist' displays of masculinity. These charges of indecency prompted the New South Wales Bathing Act 1894, introduced so that people could 'stand and watch the bathers without any sense of indecency or loss of propriety' (New South Wales Legislative Council 1894, 1429, cited in White 2007). The new by-laws ushered in by the Act made it obligatory for bathers to wear costumes covering the body from the neck to the knee, prevented undressing in public view and prohibited men and women from mixing on the beaches. In other cases, authorities have excluded men from public spaces, with the creation of women-only beaches and swimming pools being justified on the basis that women should be able to go swimming free from unwanted attention (see Iveson 2007 on McIvers Pool, Sydney; and Watson 2006 on the regulation of Hampstead Pond, London).

But even if women have traditionally been offered some protection from the sight of male nudity, as well as refuge from the male gaze when their own semi-clothed state is legitimate, this does not means that women enjoy the same rights as men to remove their clothing in public places. Because of the association of breasts with sexuality, women who decide to go topless in public places can find themselves subject to laws controlling lewd or indecent display, in contrast to men whose right to go topless is rarely disputed (Boso 2009). Some feminist campaigners have highlighted this inequality, with 'topfree' protests in public parks or on the streets having drawn attention to the dual standards around male and female undress. In 1986, for example, seven women were arrested for violating New York's Penal Law 245.01 ('exposure of a person') when they bared 'that portion of the breast which is below the top of the areola' in a public park in Rochester, New York.

Ultimately, however, they were acquitted when the statute was declared discriminatory for defining nudity differently for women and men. Likewise, Gwen Jacob, a philosophy and women's studies student charged in 1996 under section 137 of the criminal code of Canada was acquitted for her topfree protest as it was considered it had caused no harm: Valverde (1999) argues Jacob was classified as a rights-seeking, disembodied subject rather than an 'objectified' woman whose breasts could be regarded in any way sexually provocative.

Despite such successful challenges to legal assumptions about the indecency of exposed female breasts, in many jurisdictions women going topless have been described as reckless and causing a 'common law' breach of the peace. US courts have ruled that zoning ordinances, lewdness statutes, obscenity laws and regulations aimed at banning nudity can all be justified if it is in the governmental interest to protect public order and minimize potential harm. Assumptions thus remain that a woman who, of her own volition, publicly exposes her breasts is shameful and immoral. The principal exception to this is in the context of breastfeeding, which is assumed to be thoroughly un-erotic. Studies of breastfeeding mothers have nonetheless shown that worries about breastfeeding in public remain widespread, with many women not breastfeeding in public view for fear of offending others. This 'self-censorship' can be a factor in the early discontinuation of breastfeeding because of the impossibility of breastfeeding successfully without doing it in public (Acker 2009). The act of exposing one's breasts in the course of feeding an infant stands on the threshold between acceptability and disavowal, raising serious questions about the boundaries of private and public:

> Although breastfeeding is very often a private domestic event, the problem remains that, at times, it necessarily has to occur in public space . . . there are various taboos and spatial rituals associated with breastfeeding, which serve to position breastfeeding space as liminal space. These demarcated spaces and places are used as 'transitional zones' which women move in to breastfeed and out of to reintegrate with society and 'normal' daily activities.
>
> (Mahon-Daly and Andrews 2002, 70)

Even within the semi-private spaces of clinics or public toilets, Mahon-Daly found that women seek to develop spaces of privacy for what they regarded as an intimate act, with breastfeeding in view of men being regarded as particularly embarrassing.

Partly because of this sense of prohibition, public nudity of any type has the potential to be not only abject and deviant but also exhilarating. This idea of nakedness as liberating and life-affirming is central to the discourses of naturists and nudists, who typically justify their practices as a source of personal relaxation, freedom or esteem which involves a new way of seeing and being seen (Barcan 2001). Alongside the idea of nakedness allowing participants to escape existing categories of status (Holmes 2006), there remains an important emphasis on developing a distinctively embodied and ethical relationship with nature. Despite claims by naturists that to be undressed is to be 'naked as nature intended' (Bell and Holliday 2000; Obrador-Pons 2007), naturism courts controversy, and remains prominent in media debates concerning sexuality, morality and civility (Winship 2000). In such debates, ideas that nudity is pure, natural and healthy collide with ideas that it is inappropriately sexualized and anti-social:

> Naturism occupies a paradoxical position in western society. Advocates have celebrated it as *the* authentic human–nature relationship, a way of re-kindling our connections with the natural world, and a means of achieving and maintaining physical, mental and spiritual health. Yet, naturists have been, and still are, frequently vilified by the press, dismissed as morally ambiguous cranks, and satirized by wider society.
>
> (Morris 2009, 283)

While Obrador-Pons (2007) argues that nudism is first and foremost an expressive and affectual practice – 'a way of accessing the world through the body and a sensual disposition' – the social taboos surrounding nakedness force naturists to manage their appearance and dress carefully according to location. In the UK study conducted by Smith and King (2009), most of the thirty-nine self-identified naturists interviewed expressed a wish to go naked in urban and rural public environments without the fear of arrest, social ridicule or, in the case of several women, the fear of being raped. A key theme in the study was that both male and female naturists felt a need to present themselves in ways that made it clear that their nakedness was not connected to sexuality, avoiding the possibility of appearing 'perverted' or predatory. For some, even covert nudity was problematic for fear it would be viewed as suspect:

> I would do it [be nude] in quiet places and if I got caught [by a member of the public] in a situation like that I wouldn't immediately try and cover up simply because that implies I'm doing something I'm ashamed of. So you've got to balance the two together.
>
> (Craig, 50s, cited in Smith and King 2009, 443)

In this sense, naturists tend to argue that their decision to be undressed is not in any sense sexual. Moreover, in commercial naturist environments, such as 'nudist camps', sexual behaviour tends to be strictly policed so as to reproduce that environment as essentially asexual. Against this, however, some participants in Smith and King (2009) suggested that their nakedness provided the basis for an exploration of sexual feelings. In less tightly regulated nudist areas, typically beaches, naturists reported that social nudity was often connected to feelings of sexual arousal. Reported sexual behaviours in such environments, while inadmissible for some naturists, suggest that nudist spaces can be emancipatory spaces where the combination of anonymity, bodily nakedness and freedom produces what Andriotis (2010) refers to as an 'erotic oasis'.

This given, it is unsurprising that many naturist beaches are subject to police surveillance and official intervention to eliminate any sign of sexual behaviour, with media discourse continuing to figure naturists as potential sex criminals. For example, Booth (1997) charts the formation of protest groups in Sydney in the 1970s opposing the designation of Lady's Bay (Watson's Bay) and Reef Beach (Balgowlah Heights) as nudist areas, noting exclusionary discourses stressing that this designation would attract 'sexual deviants' who would prey on local children. In such cases, the idea that nudity was the prerogative of gay men was emphasized, perpetuating particular stereotypes of naked male embodiment as sexualized and reinforcing tropes that associate nudity with sexual danger.

Sexing the streets: how can public sex become a source of pride?

So far this chapter has discussed the problematic exposure of the erotic body. However, the punishment or sanctioning of inappropriately displayed nudity relies on the naked body being observed. This act of being observed is of course an ambivalent one, and it is clear that many of those charged with surveying space may find voyeuristic pleasure in gazing upon naked bodies. This is particularly the case when they can gaze but remain unobserved themselves (as in Foucault's original conceptualization of *panoptic* surveillance as unverifiable but all-seeing). Bell (2009b) considers the significant connection between surveillance and sex, citing both the voyeuristic deployment of CCTV cameras by their operators (see Koskela 2004) and the burgeoning availability of reality porn on the Internet (see Paasonen 2010) to suggest that taking erotic pleasure in being seen can be a powerful form of resistance in societies that are heteronormative and structured around the male gaze.

Whether or not some of the genres of reality porn currently prevalent on the Internet (including covert videos of 'public violation', upskirt-porn and voyeur porn) are resistive to sexual and gender norms is questionable given much content appears to be posted without consent having been given. Yet there is little doubt that public nudity is an effective tool of protest, especially in societies where there is a legislated taboo on public nudity (Barcan 2001). This has been evident in any number of protests where campaigners disrobe not to argue for the right to nudity, but to draw attention to other causes. This is particularly evident in protests organized by women's groups (e.g. see Sutton 2007 on nudity at the 2005 World Social Forum; Cresswell 1994 on 1980s women's peace camps; and Kutz-Flamenbaum 2007 on nudity in contemporary anti-war protests). Sutton (2007) argues this is a useful tactic because the presentation of women's nakedness on their own terms and for their own political ends disrupts dominant notions which allow for female nudity only in private contexts where women can be sexually objectified. Furthermore, Sutton (2007) suggests it takes 'emotion and intimacy' out of the private realm and inserts it in a public realm that is understood to be the realm of rational debate. However, the publicity given to naked protests by women can imply that disrobing is the only means of political expression available to them. Moreover, it leaves a nagging doubt that spectators are not listening to the arguments of these women, merely exercising a voyeuristic gaze over their bodies (Alaimo 2010) (see Figure 4.1).

Despite such dangers of voyeurism, naked protests demonstrate that the sexualization of public space has considerable power to challenge dominant ideologies by disrupting expectations of what is normal in public space. This is a theme that has been emphasized in literatures on gay and lesbian experiences of urban space, with numerous studies suggesting that the heteronormativity of the streets can be punctured through visible performances of homosexual identity. In this sense, the enactment and display of non-heterosexual sexualities potentially 'destabilizes' the heterosexual norm, 'constituting queer space through transgressive practices and gender performances' (Eves 2004, 492). Often these performances do not have to be particularly overt to expose the heteronormativity of public spaces. For example, Valentine (1996) cites the example of two lesbian identified women being thrown out of a British supermarket for kissing to underline that expressions of an 'authentic' gay and lesbian identity remain 'ghettoized' to the private spaces of the bar or the space of the home. While the sight of a heterosexually identified couple kissing in public space would rarely attract attention or comment, its performative repetition has constructed the idea that this is natural and normal. In contrast, the sight of two men or two women kissing remains transgressive in the sense that it disrupts the 'visual, emotional, moral and political, fields of heteronormative expectation' (Morris and Sloop 2006, 24).

Figure 4.1 Anti-vivisection protest, Barcelona, 2006 (photo: Creative Commons/Jaume Ventura).

The potentially transgressive power of 'queer' public performance needs to be understood in the context of long-standing histories of homophobia and even 'gay-bashing' in public space. As was described in Chapter One, the heterosexualization of urban space occurs through processes both subtle and overt, including self-policing by lesbians and gay men, their physical exclusion from particular spaces, the manifestation of moral disapproval and the threat or use of violence. For gay and lesbian identified individuals, negotiating the street thus presents a series of tactical choices, principally whether to 'pass' as heterosexual or whether to perform their body in ways that challenge the heternormativity of the street. Paradoxically, perhaps, this can include performances which are enactments of exaggerated versions of normative femininity and masculinity (Bell *et al.* 1994 discuss the lipstick lesbian and gay skinhead respectively). These perfomances can destabilize the regulatory fictions of heternormativity by placing in public view a disjuncture between an embodied gender appearance and assumed sexuality: the performance parodies gender and sexual norms through what Butler (1993) terms masquerade (see Chapter One). The ambivalence of such embodied performances are clear, however, as such parodies of normative identities can be misread by the observer, and can even reinforce conventional understandings of gender and sexual identity. While transgressive, the public visibility of effeminate male bodies and masculine female ones can also be misinterpreted, or provoke punitive responses (Doan 2004).

All this underlines that the performance of identity in public space is crucial in the articulation of queer politics. Valentine (1996, 151) suggests that rather than simply visibilizing gay and lesbian identities to claim a right to public space, queer represents a more far-reaching contestation of the production of public space itself. For disparate queer activist groups – for example, ACT UP, Queer Fist, Lesbian Avengers, Queer Nation – queer transgressions of public space have been perhaps the most important means of articulating an opposition to heteronormativity. Although 'gay villages' can visibilize gay and lesbian lifestyles, collective actions which take queer out of these often-sequestered spaces can effectively queer mainstream space, with actions such as collective kiss-ins filling the city's streets 'with the juices of unofficial enjoyment: embarrassment, pleasure, spectacle, longing' (Berlant and Freeman 1992, 158). For US anti-assimilationist activist group, Queer Nation, kiss-ins were part and parcel of a confrontational queering of public space which challenged its implicit heterosexuality:

> Moving out from the psychological and physical safe spaces it creates, Queer Nation broadcasts the straightness of public space, and hence its explicit or implicit danger to gays. The queer body – as an agent of publicity, as a unit of self-defense, and finally as a spectacle of ecstasy – becomes the locus where mainstream culture's discipline of gay citizens is written and where the pain caused by this discipline is transformed into rage and pleasure. Using alternating strategies of menace and merriment, agents of Queer Nation have come to see and conquer places that present the danger of violence to gays and lesbians, to reterritorialize them.
>
> (Berlant and Freeman 1992, 117)

In many ways, Queer Nation's creation of temporary queer space in public view draws on the logics of carnival, with the 'serious' politics of queer activism taking playful and pleasurable forms as participants articulate a subversive sexual identity (merriment as well as menace). Bakhtin's (1984) concept of the *carnivalesque* is relevant here in so much that public enactments of queer politics invert the assumptions that adhere to everyday space. In Bakhtin's classic formulation, carnival represents a reversal of normal social orders, so that high cultures and mindful activities are replaced by emphasis on 'low' cultures, hedonism and bodily pleasures. Carnivalesque events thus represent a temporary liberation from prevailing social norms, and a suspension of rank, privileges, norms and prohibitions.

Lewis and Pile's (1996) study of the Rio Carnival suggests that the performance of indeterminate gender and sexual identities (via transvestism) has long been part of carnivalesque inversions and mocking of everyday order. In the context of the Rio Carnival, however, they suggest that the 'policing' of the performances has rendered male performances of idealized femininity less of an inversion of everyday norms than a process that allows excessive femininities to be normalized. In this setting, drag performances may be ineffectual as a form of resistance because the performance of idealized femininity by the 'male' performer is too convincing to be read as anything but female. In other cases, however, cross-dressing and drag can be read as masquerade because of the context of viewing, with such carnivalesque behaviour being especially visible in the context of Lesbian and Gay Pride events (Browne 2007). Though many activists have traditionally argued that public hostility to lesbian, gay and bisexual identities might be diluted if questions of love were emphasized over sex, the Gay Pride movement emerged in the 1980s and 1990s to present a more overtly queer case for the acceptance of lesbian and gay sexualities on their own terms. Pride events like Sydney's Mardi Gras, London's Pride in the Park and Auckland's HERO Parade (see Case Study 4.2) have not downplayed sex, but revelled in carnivalesque displays, flamboyant costumes and floats, overt displays of same-sex desire and near-naked embodiments, all of which are integrated into a form of urban spectacle that fundamentally transforms the city, however fleetingly. Pride does not simply inscribe streets as queer: it actively produces queer streets.

Pride events have obviously been important for visibilizing queer identities. However, academic accounts of Pride are shot through with an ambivalence about their effects given such carnivalesque celebrations are inextricably linked to the regulation and disciplining of society. Pride is, after all, a licensed event, often tied in to the politics of local economic growth and designed to boost tourist spending (Hughes 2003). On this basis, Browne (2007) identifies Pride events as 'moments' whose politics are ambiguous:

> This is not to say that Pride is not political. Rather it is to contend that critical consideration of the politics and playfulness of Pride needs to explore the celebratory and commercial tensions of contemporary Prides along with the individual negotiations of these spaces. It is in these messy entanglements that Pride events, with their histories of protest marches coupled with flamboyant displays of non-normative sexualities and genders, can be located.
>
> (Browne 2007, 67)

For Browne, Pride oscillates between collective political action and communal partying, such that these boundaries become blurred. She concludes that Pride is important not just because it queers the heterosexual streets but because it

Figure 4.2 Gay Pride, London, 2009 (source: Anemone Projections/Creative Commons).

CASE STUDY 4.2

Performing Pride: Auckland's HERO parade

Pride events characteristically combine ideas of performance, entertainment and protest, consisting of widely advertised and promoted outdoor parades and associated dance parties (Markwell 2002). The first Pride event was held in New York in 1970, one year after the Stonewell riots (see Chapter Two). Many of these have become hugely popular events: London Pride in July 2010 attracted upwards of one million people, making it the UK's largest outdoor event; Amsterdam Pride in 2008 boasted around half a million; while Sydney's Gay and Lesbian Mardi Gras attracts around 800,000 spectators annually (Kates and Belk 2001; Markwell 2002).

Auckland's HERO Parade was on a smaller scale, attracting around 200,000 onlookers until its demise in 2001. HERO was a month-long festival of gay events culminating in a parade and party, strategically scheduled to occur two weeks prior to the Sydney Gay and Lesbian Mardi Gras to allow international tourists to participate in both events. Johnston (2001) documents the rise of the HERO events and their growing appeal for 'straight' audiences given both the parades and parties developed a reputation as 'well organized, well performed, and *risqué*', underlining that for some the partying may be more important than the politics.

Despite the evident appeal of the spectacle of HERO for many heterosexually identified tourists, Johnston suggests the event was opposed by some because of its blatant challenge to the heteronormativity of public space. For example, an Auckland City Councillor and Deputy Mayor (quoted in Brickell 2000) referred to the inappropriateness of some of bodily displays evident at the first HERO parade in 1994, noting there had been 'a whole lot of men that had G-strings on and nothing much else, and bare-topped women, and just a lot of sights that I don't think are suitable for Auckland'. Johnston suggests this opposition to the parade constructed the participants as deviant 'Others' accused of glamorizing a lifestyle that was 'grotesque, sleazy and not infrequently injurious to the health', while the watching tourists of Auckland were 'constructed as the "Self" or the dominant straight "mind" of Auckland' (Brickell 2000). For opponents, crucial evidence of the immorality of HERO participants was the partial nudity on display, cementing associations in their mind between obscenity, indecency and a homosexuality regarded as 'excessive'.

Charting some of the homophobic reactions to the HERO events is interesting given straight identified spectators were estimated to comprise three-quarters of the watching audience. It is also interesting to note that, according to Brickell (2000), the most 'subversive' costumes and performances (which integrated fetish wear and same-sex BDSM) did not actually disrupt or destablize the heterosexual viewer's assumptions about sexual categorizations, but reproduced a dichotomy between a heterosex seen as pure and a homosex imagined as excessive and even grotesque.

Further reading: Brickell (2000); Johnston (2001)

provides a basis for collectivities and mobilizations which do not just exist for the day of the parade, but which take shape over the course of the year, indicating, as she puts it, 'the desire to escape "364 days of crap" and the willingness to change those 364 days' (Browne 2007, 69).

But Pride has its enemies. Rejecting the idea that Pride should be the only articulation of queer identity, radical and anarchist queer groups have suggested that Pride is complicit in a commercialization and evisceration of queer desire that markets it to capitalist ends. Its representation of gay, lesbian and bisexual lifestyles is, for some, *homonormative* rather than queer. Gavin Brown (2007, 2686) explores this in the context of the queer radical group Queeruption London, who have sought to promote 'sustainable ways of socialising as queer people which are not overly mediated by the commodity'. Promoting queer identity politics rather than lesbian and gay identities per se, this group have sought to provide free, non-commercial alternatives to the commercialized and apolitical spaces of Pride and Mardi Gras: Brown (2007) describes a week-long Queeruption event in an abandoned tenement which featured homemade entertainment, performance art and sex parties, all aiming to be participatory and inclusive. In the US, Gay Shame provides a different critique of Pride by organizing events which critique the commercial sponsorship of Pride, and question the assimilationist agendas which underpin this licensed transgression of public space (see also Johnston 2007 on the importance of shame in Pride events). Such interventions suggest that public actions such as gay and lesbian parades and kiss-ins cannot always be described as radically queer (Morris and Sloop 2006).

The erotic possibilities of the public: in what ways is public sex empowering?

So far in this chapter it has been suggested that being in public implies both visibility and publicity. Yet public space can also play a converse role by providing a veil of anonymity for those seeking sexual freedoms outside the constraints of their home and working lives. Far from wanting their sexuality to be visible, such individuals pursue sexual pleasures in the public spaces of the city in a more clandestine manner, typically in spaces where they are assured that they will not be compromised by an encounter with a friend, neighbour, work colleague or even their 'regular' sexual partner. 'Public sex' is thus a form of sex pursued below the thresholds of public visibility but in publically accessible spaces (Warner 2000).

A disproportionate amount of the literature on sexuality and space has focused on male experiences of public sex environments, often drawing on auto-ethnography (Howard 1995; Tewksbury 1996; Leap 1999; Delany 2001; Douglas and Tewksbury 2008; Frankis 2009) Perhaps the most infamous study is Laud Humphreys' (1970) 'Tearoom Trade', which focused on 'impersonal' male same-sex encounters in the US. Based on covert observation methods, Humphreys took the role of a look-out (or 'watchqueen') in public sex environments to justify his presence as a participant-observer without having to disclose the purpose of his research. Given this lack of disclosure, and the fact he traced the home addresses of men he observed having sex via their car licence plates, Humphreys' work proved controversial upon publication. Nevertheless, it stands as something of a classic text given it revealed what was felt at the time to be a surprisingly high degree of sexual activity in the men's restrooms ('tearooms') of urban public parks and roadside truck-stops. Humphreys argued that the physical accessibility of these sites for large numbers of men, both with and without cars, was important for ensuring there was a continual flow of new arrivals to proposition even if 'regular' trade was not passing. But as important was that the location and design of public bathrooms provided a degree of impersonality:

> [T]hey are available and recognizable enough to attract a large volume of potential sexual partners, providing an opportunity for rapid action with a variety of men . . . such features enhance the impersonality of the sheltered interaction.
>
> (Humphreys 1970, 3)

The suggestion was the ideal tearoom was simultaneously accessible but isolated, away from the gaze of women, children and the police but close enough to a main road should a quick getaway be necessary.

Tied into practices of cruising for sex (see Chapter Six), men's use of public toilets for impromptu and anonymous sex has been a widely noted phenomenon, and has a lengthy history (see, for example, Peniston 2001 on the use of urinals for male sex in late nineteenth-century Paris, and Houlbrook 2000 on 'cottaging' in London in the inter-war years). Segregated on gender lines, public toilets provide spaces forbidden to women where men have a justifiable reason to expose themselves, with cubicles providing privacy for those who seek it. However, public toilets remain openly accessible, and for most men are used only for the purpose for which they were intended (i.e. urination and defecation). This means that a complex set of rules are adhered to so as to distinguish between those cruising for sex and those who might be offended by such approaches. This 'sexual vernacular' is predominantly non-verbal, involving a series of looks, glances and gestures that may culminate in exposure, touching, masturbation and oral sex, more rarely anal sex, sometimes in view of other men. The knowledge that such actions are going on in public toilets, principally at night, has often caused anxiety among the forces of law and order (Johnson 2007; Ashford 2008; Hennelly 2010). This anxiety has been reflected in the changing design of toilet spaces. For example, the incorporation of white porcelain and tiling in twentieth-century municipal toilets was part of a strategy for signalling those spaces were hygienic and unavailable for homosex, but Brown (2008a) argues these features came to be key parts of the sexual encounters that occurred in these settings, being eroticized through association and practice. Latterly, toilets have become increasingly gender-neutral and 'family friendly', with the incorporation of baby changing facilities and mixed aged facilities in male toilets effectively rendering them heteronormative. Attempts to make toilets more open have also been used to exclude male sexual activity, with CCTV at public toilet entrances a reminder that these are subject to the watchful gaze of the state and law (Jeyasingham 2010).

As technologies of bodily concealment, public toilets are neither fully public nor private, constituting zones whose indeterminancy comes packaged, with 'a distinctive current of psychic charge' (as Edelman 1994 puts it). Delph (1978) thus describes public toilets as 'erotic oases' for men seeking impersonal sex with men (see also Tewksbury 2010 on impromptu sex). Other significant public sex environments include beaches, car parks, urban woodlands, gardens and wastelands. Parks, in particular, have been popular sites for cruising, especially when located on the urban fringe, given men may assume these are anonymous zones where personal information about them and their activities will not be collected (Walby 2009). The divide between night and day uses of public spaces is significant here, with cruising often occurring under the cover of darkness to protect the anonymity of participants. Crucially, this includes many men who do not identify as gay, with one of the key factors underpinning the enduring appeal of cruising being the involvement of heterosexually identified men, including those in long-term relationships. In her study of Dutch public sex environments, Bulkens (2009) estimates that some three-quarters of those frequenting cruising grounds were married, 'family' men, who used these spaces to take in a new relationship towards themselves, redefine their subjectivity and explore sexual pleasures denied to them in their long-term relationships. For such men, the clandestine nature of public sex environments can be vital given many will wish to shield their sexual experimentation from their families. This underlines that the primary aim of cruising – to negotiate and experience anonymous sex – can be a process fraught with risks. As well as risks of being exposed by the police or non-cruisers, there is also the danger of attracting homophobic violence from those who regard the use of public space for such purposes as a perversion.

The extent to which threats of exposure or arrest create a frisson of excitement for those visiting such spaces is difficult to gauge, but most qualitative studies suggest that the pleasures of cruising include this negotiation of safety and danger (Frankis 2009). This suggests that cruising grounds are liminal spaces bequeathed with certain transgressive and erotic properties. The same might be said of heterosexual practices of having sex in view of others ('dogging' in the UK, 'exhibitionisme' in France), something that typically occurs in urban fringe green spaces such as nature reserves, parks and picnic areas, mainly at night. Bell (2009a, 384) suggests dogging sites provide particular environmental affordances, 'being accessible by car (and providing the all-important parking space)', offering seclusion but being easy to find, and providing a 'stage' on which 'elective privacy and publicity' can be performed. Following Foucault, both Bulkens (2009) and Gaissad (2005) hence conclude that public sex environments constitute *heterotopias*, a term Foucault used to describe the counter-public sites that are an inversion of everyday space. For Gaissad, public sex environments fit well to Foucault's notion given they have distinct ways of 'opening and closing' that isolate them spatially while allowing for their social penetration, by those who understand the rules of engagement (see Case Study 4.3).

Characterized by what Frankis (2009) refers to as a near-universal sexual etiquette, public sex environments nonetheless take a variety of forms. Drawing on extensive ethnographic work in Marseilles, Perpignan and Toulouse, Gaissad (2005) suggests that these sites are far from fixed in space and time, and represent overlapping 'sites of circulation' through which men pass to experience discrete, anonymous sex. Given tendencies towards the repression of 'anti-social' sex (see Sanders 2009a), Gaissad (2005, 25) also notes the dispersal of public sex environments from the centre to the 'edges of what is established, organized and sedentary'. Such centripetal tendencies have resulted from efforts to 'design out' public sex environments that have been justified in the name of community safety. For example, Camden Council began attempts to stop men from cruising in Russell Square gardens in Bloomsbury in 1995, when they cut down shrubs and strung lights in the trees to minimize the spaces in which sex participants could hide. In 2002, they renegotiated the lease with landowners Bedford Estate and closed the site at night. Since there are no other parks that stay open around the clock in Central London, this effectively forced cruising out to more peripheral parts of the city. Responding, gay rights group OutRage! blamed the Council for increased complaints about sex in the Square by increasing its visibility, and suggesting complaints could have been reduced had the Council turned off the lights and replanted head-high dense bushes around the outer perimeter (see also Walby

CASE STUDY 4.3

The rules of the 'beat': regulating Melbourne's public sex environments

In Melbourne, Victoria, beaches including Black Rock and Sandridge, as well as large civic parks and gardens like Fitzroy Gardens, Mornington Park and Malvern Gardens, have been notorious spaces for male cruising and impersonal sex. For those in the know, each of these have hosted public toilets and cruising areas that became known through a coded vocabulary indicating both the design and clientele (e.g. 'The Lobster Pot', 'Stiffies' and 'The Haven'). These public sex environments – or 'beats' – were subject to periodic surveillance by the police with frequent arrests of men for 'loitering for homosexual purposes', an offence until the decriminalization of gay sex in Victoria in 1981. Since then, continuing complaints about men hanging around public parks or toilets have triggered entrapment campaigns where male police officers pose as men seeking sex, and subsequently charge men for indecent exposure or offensive behaviour. While for some men the risks of arrest may add to the excitement of the beat encounter, Iveson (2007) notes that these spaces have also attracted homophobic violence, meaning that the use of public space as a cruising ground has brought beat users into danger in a variety of ways. In the 1980s, the authorities began to suggest this included the additional risk of HIV transmission, noting that many of those who had anonymous sex with men in cruising grounds did not identify as 'gay' and were not necessarily receptive to safe sex messages.

Iveson suggests this identification of cruising grounds as areas of health risk led to exercises of surveillance in which the beats were subject to a socio-scientific gaze and 'mapped' out so as to be comprehensible (see also McGhee and Moran 2000). New sexual health outreach projects began to distribute condoms and safe sex literature in beat spaces, seeking to make these safer spaces, though potentially rendering them more visible for what they were. At the same time, changing police attitudes towards cruising meant that there was increased recognition of the legality of men meeting men for sex in Melbourne's beats: legal cases suggested that if police had to seek out indecent behaviour, then it could not be deemed to offend public decency (the same being said of sex occurring behind a locked vestibule door). Yet parks authorities and other opponents of public sex environments have, Iveson notes, continued to intervene in the design and layout of Melbourne's public toilets to make them less amenable to sexual encounter (typically locking them at night).

Iveson thus presents Melbourne's beats as clandestine spaces whose 'sexual vernacular' has been revealed in the midst of struggles to make the beats safer and legal. This 'outing' is not, he suggests, something that should be read negatively given that spaces beyond the gaze of the state and law are inherently risky and dangerous spaces. Whether or not the beat spaces continue to present a challenge to heteronormativity is a moot point though, given the increasing mainstreaming of knowledge of their practices and languages: Iveson's conclusion is that such information is itself a form of publicity that begins to stake a claim for having accessible same-sex environments that exist outside the spaces of the home or the marketplace.

Further reading: Iveson (2007)

2009 on the re-aestheticization of Ottawa's public parks and efforts to design out cruising in Canada).

Yet this dispersal of public sex environments by urban authorities is occurring with the encouragement of some lesbian and gay campaigners, who condemn what Delany (2001, 65) characterizes as the 'lubricious and lazy generosity of gay public sex' as playing into heterosexual myths of gay male promiscuity (see also Leap 1999). McGhee (2004) argues that this discouragement of public sex has come about through processes of assimilation by which lesbian and gay communities have been increasingly encouraged to exercise a self-surveillance that figures them as 'good gays'. As he describes, the bequeathing of legal rights of protection to gay and lesbian populations, and the criminalization of homophobia, are both indicative of increased citizenship rights which these populations now enjoy. In the UK, this means they have the right to enter the public realm as a queer identified individual. But these rights come with responsibilities, and it appears one of these is the requirement to refrain from sexual activities that can be deemed as anti-social (see also Johnson 2007 on the relationships between police, local authorities and gay 'communities').

Although public sex environments have been a long-standing and recalcitrant part of the urban sexscape, this suggests tendencies towards the privatization of anonymous and exploratory sex, and a move from public sex environments towards commercialized sexual premises including gay bathhouses, saunas and dedicated sex clubs and dungeons (Berube 2003). While these spaces are subject to rules of engagement similar to public sex environments, as licensed and commercial premises they charge admission fees and may have hierarchies of belonging that can be off-putting to new arrivals. Given these spaces are closed, surveyed spaces, it is hard to conceive of them as providing some sort of 'sexual commons' (Boydell *et al.* 2007), with Chisholm (1999, 76) describing them as 'microcosms of capitalism' which spectacularize the male gay body, keep out 'rough trade' and marginalizing women and transgendered individuals. Despite this, ethnographic work on sex-on-premises venues suggests that they can cater to a wider range of sexual predilections and social groups than many public sex environments, and also notes the existence of many premises which encourage the participation of women in impersonal and experimental sex, whether as part of a heterosexual couple (swinging clubs), as a participant in BDSM (fetish clubs) or as a pro-sex, erotic-positive feminist (lesbian bathhouses) (Worthington 2005; Holmes *et al.* 2007; Nash and Bain 2007; Stryker 2008a; Hammers 2009). Though not public in the conventional, topological sense, such spaces enact a form of resistance by confronting the limits of what is deemed legal, moral and ethical, developing codes of conduct which might be rolled out into spaces beyond. Referring to such spaces as productive and transformative, Stryker (2008a) concludes her reflections on transgendered BDSM venues in San Francisco by suggesting that sex-on-premises venues 'set bodies in motion', with traces of these generative locations adhering to the 'mobile architecture' of the body as it extends into the world outside.

Conclusion

Houlbrook (2006) argues that one key trajectory in the history of sexuality since the eighteenth century has been the gradual privatization of sex, and a broad movement away from public manifestations of sexuality. For Berlant and Warner (1998, 24) this privatization of sex, and the concurrent sexualization of private personhood, has been fundamental in the production of the heteronormative, making sexuality seem 'like a property of subjectivity rather than a publicly or counterpublicly accessible culture'. From a queer perspective, it might be argued that the maintenance of the liberal public–private distinction privileges heteronormativity by insisting that all sexual conduct is a private matter, denying other versions of sexual expression any sort of public life and normalizing a ubiquitous and mundane heterosexuality (Hubbard 2008). The idea that homosexuality, bisexuality, non-monogamy and other 'perverse' sexualities are only tolerable if they remain in private is thus normalized through the division of urban space into public and private, where the former is deemed de-sexed or unsexy yet is steeped in heteronormativity.

While the tendency towards the de-sexualization of public sex seems to hold true in many contexts, to assert such a general tendency is to gloss over exceptions to this rule, as well as to fail to register the shifting understandings of public and private that abound in different times and spaces. Moreover, the public–private distinction itself becomes unsustainable in the midst of many sexualized acts, suggesting that a more fluid interpretation of public and private space is necessary to understand the changing geographies of sexuality than one based, for example, on topological distinctions between home and street. Public sex environments, for example, offer degrees of privacy and seclusion that allow for sexual experimentation, while some spaces that are nominally private can be effective spaces from which to generate publicity.

Even so, the distinction between public and private is one that remains widely fussed over in discussions of sexual propriety and bodily comportment, with instances of perceived transgression retaining their capacity to shock and disgust, especially where the site of transgression is highly visible:

> Transgression, sexual or otherwise, in iconic places of 'propriety' and visibility is a particular problem for governance, we suggest, because it frequently prompts particularly vociferous and sometimes violent reactions, since it is regarded to be an affront to 'common values.' Visible transgression, such as the occurrence or representation of Other sexual practices and identities, destabilizes hegemonic norms while state reactions to those disruptions – efforts to actively shore up or adjudicate the space's 'natural' order – reveal that order as social, power-laden, and uneven.
>
> (Catungal and McCann 2010, 90)

This means the visible eroticization of the city has considerable power to challenge taken-for-granted assumptions about social and sexual order, constituting a boundary-crossing that attracts moral condemnation and flashpoints of legal contestation. Transgressive public sex is hence a modality of action in which 'the body communicates sensually and emotively with others to produce transformative attitudes' (Hastings and Magowan 2010, 11). It is this logic that is seized upon by sexual outlaws and dissidents who have sought to make their sexualities visible on the streets in transgressive moments such as Pride.

As this chapter has shown, there is much discussion as to whether making other sexualities visible in public space, or queering the streets, is an effective political strategy for destablizing existing and taken-for-granted assumptions about sexuality. Much clearly depends on context, and who is, or isn't, looking. Developing this argument, Ravenscroft and Gilchrist (2009) argue that public sex environments are simply the spaces where individuals are allowed to do things that would be 'out of the question' in other times and in other spaces. Most public sex environments and dogging sites, they argue, are returned to 'normal' during the day, being sites of family leisure rather than anonymous sex (something that is acknowledged in public sex environment etiquette, which demands that no signs of illicit activity are left). These sites are therefore subject to control, and cannot be considered as necessarily liberatory.

Bev Skeggs has therefore written of public visibility constituting a 'trap' for sexual dissidents, suggesting that the transgressive potential of public sex is often limited:

> It summons surveillance and the law, it provokes voyeurism, fetishism, the colonist/imperial appetite for possession . . . it reduces the body to the sign of identity . . . Only some groups can positively and resourcefully spatialise the claim for recognition via visibility . . . and only some groups can legitimate and/or symbolically convert their visible claims.
>
> (Skeggs 1999, 228)

In a somewhat similar manner to those commentators who suggest that the public acceptance of gay villages is, more correctly, a strategic appropriation and commodification of the 'pink pound' (e.g. Knopp 1995), Skeggs argues that public acceptance and recognition of dissidents inevitably relies on them accepting certain compromises. For example, this chapter has described how, in asserting their claims to equal citizenship, lesbians and gay men mark off their bodies as different from the heterosexual 'norm' through forms of drag and heterosexual parody. For many sexual dissidents, the type of gaze this attracts, and the attitudes it engenders, may be problematic given the homophobic intolerance that persists in many cities (e.g. see He 2007, on Taipei's Gay Pride). Even in a society that grants them full citizenship, when brought into public view sexual dissidents may become all too visible, vulnerable to any backlash that the putative mainstream might later unleash. Sometimes, claiming the right to sexual privacy appears more important than fighting for the right to occupy public space on others' terms.

Further reading

Barcan (2004) is a rich and diverse history of nudity that considers its shifting cultural meanings and considers the development of prohibitions around the sexualized, naked body.

Califia (1994) represents an important intervention in queer studies by identifying the possibilities and limitations of transgressive sex, public and otherwise.

Johnston (2005) is a geographical study of lesbian and gay tourism that focuses on the importance of Pride events and their resonances for both participants and spectators.

5 On the town

Pleasure and leisure in the nocturnal city

Learning objectives

- **To appreciate that urban consumption and leisure is, and always has been, sexualized.**
- **To gain an understanding of the moral anxieties and concerns that adhere to spaces of night-life.**
- **To become aware of the ways that leisure spaces have been crucial in the definition of sexed and gendered identities.**

Studies of people's routines suggest that, after sleep, more time per day is spent in leisure than any other non-work activity, including housework, caring or body maintenance. Moreover, after housing, spending on leisure is the single biggest financial outgoing for most households. Despite flexibilization in working hours, most of this leisure spend occurs at night, away from the routines of the 'working day'. While one-third of this expenditure is on in-home entertainment, including watching TV and DVDs, surfing the Internet, playing computer games and listening to music, the majority of leisure spend occurs outside the home. Indeed, one of the characteristics of urban life is that the city provides a particularly dense concentration of public amusements, including theatres, restaurants, clubs, pubs, cafés, parks and cinemas, and that these spaces provide the focus for rituals of night-life and leisure. While the popularity of these different spaces of night-life has waxed and waned, with a major shift towards 'out of town' leisure noted in the 1980s and 1990s (Watt 1998; Hubbard 2003), the city centre remains the principal focus of urban night-life, in many cases being aggressively marketed as an urban 'playscape' and an entertainment 'hotspot' as the evening economy has moved centre-stage in discourses of economic growth and revitalization (Chatterton and Hollands 2003).

While contemporary 'post-industrial' urban night-life is often described as divided on lines of class, gender and ethnicity (Zukin 1998), its sexual dimensions are more rarely explored. This is surprising given night-life is profoundly embodied, and, by definition, involves people meeting and mingling with others, experiencing diverse sensations of desire and disgust. Beyond noting the emergence of adult entertainment (see Chapter Six), or commenting on the emergence of gay venues, circuit parties and club nights (Weightman 1980; Slavin 2004; Westhaver 2005), there has been little attempt in leisure studies or sexuality research to expose the sexual practices and performances associated with spaces of night-life. As such, this chapter explores the ways that sexual identities are made and re-made in 'mainstream' spaces of night-life, and highlights the ways that these spaces are implicated in the making of normative sexual relations and the promotion of heterosexual coupledom.

Emphasizing the liminal and even carnivalesque nature of urban night-life, and its association with hedonism and 'excess', this chapter accordingly explores some of the more important sexual dimensions of urban night-life. It begins by considering the rich literature on male *flâneurialism* and rambling in urban social spaces, noting how spaces of night-life have become colonized by particular homosocialities and performances of masculinity, including forms of aggressive heterosexuality that have been considered threatening to women. The chapter thus devotes much attention to the *visual cultures* of the night-time city, and considers how bodies are put on display in different sites of urban consumption. However, it moves beyond well-trodden debates on the male gaze (and related practices of 'cruising') to consider the wider range of ways in which leisure spaces have been incorporated into rituals of 'courtship' and coupling. Noting the changing gendered dynamics of such rituals, the chapter explores the ways that women have sought to appropriate, transform and challenge male dominance of the city at night through their use and occupation of night-time leisure spaces. Noting recent media outcry concerning the behaviour of young women in British towns and cities (and the rise of the 'ladette'), the chapter concludes by considering whether sexual norms might be changing to accommodate sexually predatory feminine identities alongside male ones.

Bright lights, big city: in what ways is night-life sexualized?

Before the nineteenth century, most cities moved to the rhythms of nature, being tied into the diurnal cycles and day and night. 'Clock time' was in some ways an irrelevance given people worked during the hours of daylight, and retired to their homes at night. Oil lamps and candles provided a modicum of light, but were not enough to dispel fears of the nocturnal city. In many cities, curfews were enforced, partly on the assumption that those venturing out under the cover of darkness were criminals or vagabonds. This was to change, however, as new innovations in the production of coal gas allowed for experiments in city-wide street lighting in the early years of the nineteenth century. The first public street lighting was demonstrated in Pall Mall, London in 1807 and five years later the London and Westminster Gas Light and Coke company was founded as the first of its kind (Schivelbusch 1988; Schlör 1998).

Within the space of a few decades, street lighting became common in most towns and cities, most evident in the dazzling thoroughfares and grand urban boulevards, but extending out into the darkest recesses and rookeries of the city, gradually banishing much of the fear and anxiety associated with the city of the 'dreadful night'. Yet Schlör (1998, 58) argues that the production of these public topographies of light was not pure municipal benevolence and civic paternalism, but represented a 'comprehensive claim to power'. In the act of lighting the city, urban governors effectively made a claim to that city, stressing their ownership and control of its public spaces. Schivelbusch's discussion of public lighting systems in Paris in the nineteenth century underlines this point, suggesting that state-owned and operated public lighting allowed for round-the-clock surveillance of the city's spaces, and, in the process, became 'closely associated . . . with the repressive function of the police' (Schivelbusch 1988, 62). The disciplinary light of the street lamps thus took on a symbolic meaning of law and order: as Schivelbusch points out, many street lights were destroyed by the 'lower orders' of the Parisian populace in the 1830 and 1848 revolutionary uprisings, with lantern smashing becoming a cultural ritual signifying rebellion and resistance (see also Cresswell 1998).

Despite the disciplinary logics underpinning its introduction, there is no doubt that progressive illumination opened up the nocturnal city to a wider range of uses, and enhanced the circulation of goods and people. Moreover, when coupled with new innovations in street paving, cleansing and ordering, it made the city at night more commodious and civilized, encouraging even 'respectable' populations to venture into the city at night. Entrepreneurs were keen to take advantage, providing new attractions. An early example of this was provided by a precursor of the modern-day amusement park: the pleasure gardens. While such pleasure gardens dated back to the seventeenth century as landscaped parks they became increasingly popular in nineteenth-century London as gas lighting made it possible for their owners to offer evening entertainments such as dancing and dining. Cremorne Pleasure Gardens was one example, offering twelve acres of attractions to the paying public, including bandstands, private supper boxes and lit promenades (see Figure 5.1). As Nead (1997, 108) describes, this blend of rural retreat and theme park 'opened up the other side of daytime London: a dreamworld of sexuality and pleasure, phantoms and superstition'. While this proved immensely popular, it began to attract criticism, with contemporary newspapers reporting 'crowds of noisy, disorderly inhabitants, forming an accumulation of vice and immorality which shocked respectable inhabitants' (cited in the *Chelsea News and Advertiser* 7 October 1871). Unlike the existing pubs and taverns that characterized urban night-life as essentially male, the pleasure gardens were seen to provide new opportunities for flirtation and sociality between men and women, raising anxieties about the possibilities of inappropriate behaviour fuelled by drunkenness (see also Dreher 1997). Opponents went so far as to describe them as spaces of corruption, where young men and women might be seduced by the 'superficial charms' and distractions of night-life.

The illumination of the city went hand-in-hand with the invention of a bewildering range of new leisure opportunities, and ultimately gave birth to night-life as it is currently understood: a set of socialities played out during the hours of darkness, enabled by, and symbolized through, artificial lighting. In Paris, for example, the new gas lighting was said to dazzle the onlooker, becoming 'an essential prop in the staging of the dreamworld of urban modernity' (Gunn 2002, 26). The subsequent illumination of the grand boulevards created by Haussmann in the 1850s not only gave rise to its description as *La Ville Lumière* ('the City of Light'), but led to a number of technological and cultural innovations (e.g. the pavement café, the cinema, the revue bar, the billboard, the news-stand) which were tied into rituals of night-walking, 'window shopping' and consumption. These Parisian innovations were mirrored elsewhere, inflected by the national tastes and moralities: Ward (2001), for example, notes the extraordinary 'frenzy of light' that transformed Weimar Berlin, describing the importance of coloured electrified advertising in making streets that appeared quite old-fashioned during the day appear shockingly modern at night. Connecting these new visual displays of night-life to the pleasures of looking (*scopophilia*), Ward concludes that the pedestrian became literally intoxicated by the sights and sounds of night-time Berlin. In New York, too, electric lighting was implicated in the making of a variety of new leisure spaces, from the cheap amusement parks and dance halls of Coney Island, through to the sights of Broadway – the theatre district and 'great white way' whose illumination spectacularized Manhattan night-

Figure 5.1 Cremorne Pleasure Gardens, *Illustrated London News*, January 1851.

life and produced an 'aesthetic of astonishment'. As McQuire (2005, 225) describes, 'electric lighting, with its unprecedented intensity, precision, and automated control set in motion a complex psychogeography of seeing and being seen' that became integral to New York's 'cityscape of promiscuous display and everyday voyeurism'. New York became a place to see and in which to be seen.

Night-walking – a leisurely amalgam of strolling, loitering and, importantly, gazing at the urban spectacle – thus became a characteristic ritual in newly electrified metropolitan centres. In Paris, such practices were famously documented in the poems of Charles Baudelaire and later in the Arcades project of Walter Benjamin. Night-walking was also indelibly associated with the emergence of a new urban 'type' – the *flâneur*. Though often posited as an ideal or metaphorical figure, the *flâneur* had a real historical existence as an aimless, complacent, wealthy member of the bourgeois class who wandered the city in search of distraction:

> The street becomes a dwelling for the flâneur; he is as much at home among the facades of houses as a citizen is in his four walls. The walls are the

> desk against which he presses his notebooks; news-stands are his libraries and the terraces of cafés are the balconies from which he looks down on his household after his work is done.
>
> (Walter Benjamin, quoted in Buck-Morss 1991, 43)

Such 'botanising on the asphalt' was only possible in the metropolis, where modern urban public places provided the ideal spaces for seeing (and being seen): Kramer (1988, 30) wrote that the Parisian streets were 'a permanent parade in which every visitor and every resident participated at one time or another'. Indeed, following the electrical illumination of the city (in the *Belle Époque* of the 1870s) it no longer made sense to talk of the *flâneur* as a purely upper class figure, for the city provided all classes the opportunity of being a *boulevardier*, whether they were the bohemian dandies of Montmartre, the haute bourgeois of Neuilly or the working classes of Belleville. *Flâneurialism*, strolling and 'rambling' was encouraged in Paris – and other major cities – by a bewildering range of guides which mapped out the 'right' spaces of the metropolis to be seen in, noting where one could experience the new, the unfamiliar and, importantly, the erotic (Rendell 1998; Howell 2001; Cohen *et al.* 2008). Such guides also advised on appropriate styles of grooming and dressing, with the act of preparing for a perambulation of the city vital in confirming the urban masculinity, heterosexuality and wealth of the 'man about town' (see Case Study 5.1).

Far from typifying the experience of modernity's public places, feminist scholars have been at pains to highlight that the *flâneur* encapsulated a mobile, subjective gaze that was profoundly male. From a feminist perspective, the non-existence of the *flâneuse* symbolizes women's restricted participation in public places as well as the gender bias in some of the classical literature on modern cities (Wolff 1985). The invisibility of the *flâneuse* within historical accounts underscores how the freedom to roam at night was very much a male freedom: the *flâneur*'s licence to watch the city sights being an obvious manifestation of his male privilege and worldliness. In contrast, Wolff argues, bourgeois women were consigned to the private sphere of the home and family and would have been judged as disreputable or immoral – typically mistaken for the 'fallen woman' or prostitute (see Chapter Two). Indeed, the 'extroverted nature' of the modernizing city, with its brash night-life attractions, effectively created the prostitute as spectacle, with Corbin (1990, 205) noting that 'she paraded herself under the bright artificial light of Paris by night, whose ostentation now stimulated the fantasies that sprang from that milieu'. The prostitute was an object of fascination for the man of the streets, who described her as a figure of both desire and disgust, being spatially illegible and sexually transgressive.

The nature of the relationship between the *flâneur* and the prostitute – much commented on in histories of the modern city – makes the point that women were an important part of the urban scene, but were only allowed to occupy the night-time city on male terms, as Rendell argues:

> Women are exchanged, both socially and symbolically, as commodities . . . men organise and display their activities of exchange and consumption, including the desiring, choosing, purchasing and consuming of female commodities, for others to look at in public space.
>
> (Rendell 2002, 19)

The idea that the modern city provided a patchwork of sexual opportunities for the 'man about town', and that men could pursue different pleasures in different leisure spaces, was hence a modern conceit, with Wilson (1995, 72) noting the importance of the young 'fertile' woman as a figure of 'public pleasure' in the masculine landscape of modernity. In this sense, certain women were allowed to circulate within the nocturnal city, their mobility understood as both a challenge and provocation to the man about town, who drew pleasure from the social and sexual confusion this created.

In most cases, the sex being sought by the dandy, rambler or *flâneur* was assumed to be heterosexual, with the male explorer seeking out the female urban 'temptress'. Yet queer scholarship notes there are certainly similarities between the *flâneur*'s pleasure-seeking strolls through the city streets and the act of homoerotic cruising. While cruising is often associated with public parks and toilets, which provide certain levels of anonymity in the hours of darkness and can often be transformed into spaces of sexual encounter (see Chapter Four), the city's streets constitute a key site of cruising activity. Indeed, practices of cruising often entail a subtle subversion of the 'normal' choreographies of street-life, whereby strangers will often display *civil inattention*, glancing at one another so as to ascertain that the other does not pose a threat, but not staring given this might appear threatening or disconcerting. Cruising relies on visual exchange, and a series of subtle signs being made in the initial moment that a body passes another body in the midst of the urban crowd: a glance that lingers just a little too long, a smile, a nod of the head. The title of Turner's (2003) study of the cultural histories of cruising in New York and London – *Backward Glances* – emphasizes the importance of looking back, of returning the gaze and of confirming that which appeared intended was in fact intended. This complex choreography of postures, movements and gestures can lead to the (unspoken) negotiation and transaction of public and anonymous sex, or in some instances, simply an exchange of numbers and an agreement to meet for a drink some time (Brown 2008a).

CASE STUDY 5.1

Consuming the urban spectacle: rambling in the Burlington Arcade, London

In the post-industrial era, shopping centres or malls have been defining spaces of leisure: neither fully public nor private, regulated through omnipresent CCTV and security, and cut off from the mundanities of everyday life, these spaces organize a disorientating range of commodities within a spectacular and seductive setting. Often, the origins of such spaces are traced back to the late eighteenth- and early nineteenth-century arcade, an architectural form most associated with Paris, where over fifty were built between 1790 and 1850, mainly on the right bank of the Seine. These covered spaces offered a traffic-free space liberated from the dirt, noise and weather that dissuaded the middle classes from venturing onto the streets, and used new forms of selling and display to fetishize and spectacularize the 'latest' commodities being produced in the industrializing metropolis. As such, the arcade was a pioneering space of consumption that transformed shopping into a leisure activity.

As one of modernity's original spaces, the arcades or passages have attracted much retrospective attention, most famously from the German philosopher Walter Benjamin, who devoted much of his life to exploring the aesthetic forms associated with industrial capitalism. Benjamin's Arcades project (*Passagenwerk*), which remained uncompleted on his death in 1940, took the arcade as an emblematic form, suggesting that it was the space where industrial capitalism's effects were most sharply evident in the nineteenth-century city (Buck-Morss 1991). Meandering through the contemporary arcades, Benjamin seized upon sights and relics of the nineteenth century, taking these to be *dialectical images* that could reveal a historic understanding in the present. Writing a history of things, not people, Benjamin's account sought to demonstrate how the 'phantasmagoria' of modernity, presented in exaggerated form in the spectacular spaces of the mall, produced a commodity fetishism that displaced any notion of exchange or use value. Things were no longer valued for what they did, or what they were worth, but for how new or fashionable they were. In this 'marketplace of desire', sex itself was one of the commodities that was exchanged in the space of the arcade, either by association or directly in the form of prostitution.

While the Parisian arcades have been subject to considerable scrutiny, they were characteristic of many other cities in Europe and beyond (for example, the US's first arcade was built in Providence, Rhode Island, in 1828). In London, the Burlington Arcade, situated between Piccadilly and Bond Street, was opened in 1819, and remains an upmarket retail setting, policed by 'beadles' who retain the traditional frockcoat and top-hat uniform first adopted in the nineteenth century. Built on land owned by Lord Cavendish, adjacent to Burlington House, the arcade represented a revolutionary public space within the capital, providing Londoners with an opportunity to promenade and view while being protected from the weather and hubbub of the outside world (Rendell 1997). A single, top-lit walkway, the Burlington Arcade offered seventy-two two-storey shops, among them eight milliners, eight hosiers or glovers, five linen shops, four shoemakers, three hairdressers, three jewellers and numerous lace-makers, booksellers, hatters, umbrella or stick sellers, wine merchants, case-makers, goldsmiths, tobacconists, engravers and florists.

Like the 'bazaars' that were beginning to crop up elsewhere in the city, contemporary accounts of the Burlington Arcade suggest it represented a magical space of enchantment, 'a site of intoxication and desire' (Rendell 1998, 9). While the enticing display of commodities drew in the fashionable and respectable classes, the effective removal of the arcade from the space of the everyday city, and its association with commodity consumption, led it to be connected with both male sexual pursuit and female display. Notably, while vagrancy laws were used to prevent soliciting on the streets, the 'private' street of the Burlington Arcade provided a place where it was possible for prostitutes to solicit in comfort without fear of being arrested. Writing in 1862 Henry Mayhew remarked on the use of upper chambers over 'a friendly bonnet shop' for purposes of prostitution. More generally, from the perspective of the male viewer, the visibility of unaccompanied women within the Arcade also implied their sexual availability, whether or not they were prostitutes. As Rendell (1998, 40) describes, the sexual excitement of the arcade for the 'man about town' lay in the presence of the 'feminine' as a 'screen for projecting fantasy': the women in the arcade were viewed as both chaste and lewd, and it was 'the rambler's inability to decipher the "true" sexual identity of a woman from her appearance which titillated'.

Further reading: Rendell (1997)

Cruising is a practice that thus exploits the ambiguities of modern urban life, and particularly 'the uncertainties that linger in the fleeting experience of a backward glance' (Turner 2003, 66). It can also be a practice that relies upon the eroticization of the street at night, with the gay *flâneur* being most certain of the meaning of the backward glance in the spaces of the nocturnal city, outside the routines of work-time. Importantly, this is the time when bodies are most likely to be performed in ways that are likely to be read as signalling particular sexualities or orientations, with Harris (1997, 35) insisting that bodies are most obviously styled and fashioned to appear as (variously) gay, lesbian or, conversely, straight in the context of urban leisure and night-life. Tellingly, however, and despite the tentative identification of the lesbian *flâneur* (Munt 1995), cruising remains a practice indelibly associated with men's search for sexual pleasures in the city.

When the sun goes down: why do spaces of night-life concern the state and law?

So far this chapter has drawn mainly on literature describing the emergence of night-life in Paris, New York, London and Berlin in the nineteenth century. However, provincial towns and cities began to develop their own evening economies, albeit often of less spectacular appearance, at the same time. By the first decade of the 1900s, for example, there was no US city without street lighting and all cities boasted vaudeville houses, nickelodeons and movie palaces whose design and lighting indicated their separation from the 'grey worlds' of home and work. These were the central institutions of the new urban night-life, their exteriors marking the city after dark as an attraction and calling attention to the delights for sale within (Nasaw 1992, 280). Zukin (1995) hence describes a fundamental transformation of the character of public spaces of US cities between 1880 and 1945 as night-life created new zones of commercial activity based on hotels, bars, saloons, theatres and cinemas. Significantly, Zukin notes that these zones were genuinely public, and that despite some owners' desire to create an air of exclusivity, spaces of night-life quickly became popular and even democratic, offering pleasures that were affordable to the lower classes as well as the bourgeoisie.

One corollary of this was that there were an increasing range of leisure spaces where women were welcome. Peiss (1987) argues that while married women remained largely absent from the public sphere, their 'leisure time' being spent on household tasks (Chapter Three), an increasing number of young, single and waged women began to experience rhythms of time and labour which had much in common with working men, and led them to expect similar access to spaces of leisure. However, overwhelmingly women's search for leisure did not lead them to the traditional haunts of working men – pubs and clubs – but rather an emergent range of commercialized recreations – dance halls, theatres, cinemas, tea shops, department stores and restaurants – which appeared to offer more respectable pleasures. In contrast, women's presence in public houses remained deeply suspect, with Kneale (1999) describing the growth of the temperance movement in the nineteenth century as strongly related to fears about the threat public houses posed to the family. Given the figure of the drunken husband was viewed as the embodiment of this threat, women in pubs were viewed as potential home-wreckers and 'harlots' – or, if young, were regarded as being subject to exploitation and corruption.

This given, it appears that at the same time as leisure was being made available for all, and the nocturnal city was apparently becoming more public, anxieties about spaces of night-life attracting the 'wrong' type of people and behaviour remained entrenched. Regulators and reformers were particularly anxious about the spaces that they felt encouraged problematic sexual behaviour by young women (Barron 1999). Indeed, while pleasure gardens had largely disappeared by the early years of the twentieth century, similar discourses alleging vulgarity and sexual immorality began to circulate around the 'cheaper amusements' that provided the focus for popular rituals of evening leisure (Peiss 1987). Film exhibition, for example, was initially described by critics as a passing, vulgar fad, but proved massively popular: the first purpose-built cinema in the UK was in Colne, Lancashire, opened in 1907, but there were 1,600 by 1910 and double that by 1918. Cinema proved especially appealing to women:

> Film content increasingly paid attention to themes such as romantic relationships and social mobility, and the film trade targeted its product on young fashion-conscious females; it was a world that seemed relevant to their concerns and one which openly attempted to include women within its boundaries. Unlike the public house, cinemas did not impose spatial restrictions on female customers. It also gave them access to a wider world beyond the confines of home, domestic duties and strictly enforced gender roles. Additionally, the cinema environment could offer a taste of exotica and luxury, and also provide an important site both for mixed and single-sex socialising.
>
> (Brader 2005, 105)

Concerns nonetheless abounded that film had the power to seduce and corrupt. This led to a careful regulation of cinemas, with the British Board of Film Censorship, founded in 1913, promising no film would be passed 'that is not clean and wholesome and absolutely above suspicion': while romance was permissible, sex was not. This control of film was accompanied by a careful surveillance of the cinema audience itself, with fire regulations justifying regular inspections of cinemas and their clientele. Single women visiting cinemas were regarded as suspect, and instances of suspected prostitution or 'treating' going on under the cover of darkness could result in revocation of a cinema licence under the terms of the 1909 Cinematograph Act. Meanwhile, in the Weimar there appeared to be anxiety that films portraying the dangers of prostitution actually mirrored the behaviour of women in the cinema, who could be 'bought' for the price of a cinema ticket or glass of wine (Smith 2010).

Identifying spaces like cinemas as sites of heterosociability and gender mixing, moral reformers accordingly argued that new forms of night-life might encourage sexual promiscuity if not suitably regulated. The fact that much of the concern centred on the possibility that young working women would flirt with working men late into the night underlines the gendered assumptions and prejudices that informed debates about the place of women in the modernizing city. This certainly came to the fore in the 1920s, when the 'flapper' emerged as a key social type (Caslin 2010). A sexually confident young woman, usually single, white and aged between sixteen and thirty, contemporary media discourse identified the flapper as an urban pleasure seeker whose appearance and comportment fundamentally challenged traditional models of femininity (Jackson and Tinkler 2007). As a quintessentially 'modern' girl, the flapper adopted fashions which were variously described as androgynous, skimpy, daring (classically, short bobbed hair, a fringed skirt, stockings rolled and bunched below the knee (see Figure 5.2)). In terms of behaviour, the flapper was thus regarded as being rather too masculine, risking her health and virtue by smoking and drinking in spaces alongside her male counterparts. Ward (2001) argues that these anxieties were profoundly sexualized, and that, by seizing some of the visual artifices of male appearance and behaviour, the flapper or new woman raised concerns about the decline of traditional notions of feminine sexuality being tied to motherhood. Symbolizing feminine liberation, the flapper was often represented as 'cheap' or 'fast', notions that had clear sexual implications, and in some accounts was even identified as complicit in the emergence of visible lesbian cultures and other 'perversions' (Messerschmidt 1987).

In many ways, the flapper embodied the new socio-sexual mobilities of the modern city, and the profound ambivalence that surrounded the increased visibility of women in spaces of night-life. On the one hand, the emancipated sexuality of women was celebrated, and put 'on display' in the night-time city, with women's bodies becoming part of the commodifed spectacle (Mort

Figure 5.2 The flapper faces moral choices, US illustration from sexual hygiene manual, *c.* 1922 (source: Creative Commons/no author identified).

2007). On the other, there was anxiety about the presence of women within the expressive commercial cultures of night-life given the persistence of the cult of domesticity and the promotion of chastity and decorum as defining feminine traits (see Chapter Three). Ultimately, however, women's presence in the commercializing, modern city was contested because it challenged the assumption that the city at night was a *virile* space that could only give up its pleasures to men. As Mort (2010, 45) notes in the context of 1950s London, twentieth-century public concern about socially mobile young women who 'drew on the resources of consumer culture as a form of empowerment' was part of a long-standing history concerning the defence of particular masculine forms of spatial privilege (including *flâneurialism* and rambling).

The idea that women do not belong in the night-time city, and are unable to handle its 'darker social and sexual secrets' (Walkowitz 1998, 87) has thus remained a persistent myth, despite the evidential importance of women as the objects of the male gaze (see Case Study 5.2). This myth has of course been challenged, and from the earliest years of the nineteenth century, campaigners argued for the right of women to go about their business free from the advances of the 'ungentlemanly'. But the desire of women to appear fashionable and of-the-moment, performing a sexualized femininity, meant that even 'respectable' middle class women were subject to unwanted male attention (Walkowitz 1998). There was accordingly plentiful advice for women as to how to dress and behave to avoid unwanted attention, particularly if they were countrywomen coming to the city for the first time (Bieri and Gerodetti 2007).

These types of example indicate that sexual harassment emerged as a significant social problem in the era when women – and especially unaccompanied ones – began to enter the city centre, traditionally a privileged space of politics and commerce, as consumers. Since that time, multiple studies have revealed the anxieties and fears that many women experience when traversing the night-time city, with the threat of sexual harassment and sexual violence impinging on women's spatial freedoms:

> Growing up in New York City in the 1930s, 40s and 50s . . . I experienced some of the great ambiguities of being a woman – particularly a young woman – growing up in a great city. On the one hand we faced an unprecedented degree of freedom: roller skates, going to parks by ourselves, going to museums, theatres, the unrestricted use of public transportation. Yet sexual harassment and even more, the threat of aberrant sexuality as an ever-present, hovering spectre, restricted our access to the public life

CASE STUDY 5.2

Objectifying women: girl-watching in Montreal

The idea that the modern city encourages an engagement with other individuals on the basis of appearance is long established in sociological literature on street life and the crowd. The city, it has been argued, encouraged a visual scrutiny of women's bodies by men that was simply not possible in the rural. In the nineteenth century, for example, new hand-held cameras were associated with 'snap shot fiends' who would prowl the streets taking candid photos of unsuspecting women (Walkowitz 1998) – a practice that is echoed in the contemporary use of mobile telephone cameras to take covert images of women in public spaces, including 'up-skirt porn' (Bell 2009a). Such behaviour is rarely considered harmless or acceptable: 'uninvited sexual attention' in the form of looks or gestures is commonly construed as sexual harassment.

Yet in some instances, and particularly prior to the advent of sexual harassment legislation, girl-watching appeared more accepted as a 'natural' part of the urban scene. Wallace (2007), for example, charts the emergence of a 'somewhat tongue-in-cheek' girl-watching movement in Montreal in the 1960s in which the media began to chart a series of ideal locations in the city where men might hope to see the most beautiful women in the city. This relied upon the city's press mapping out the sites that boasted high concentrations of (especially) young professional women, ideally dressed in the latest fashions, notably the mini skirt. Guides for the uninitiated set out the rules of etiquette – advising that the 'all-too obvious' glance needed to be avoided – and offered advice on 'where women would be before, during and after work; which streets they preferred to walk on' and even 'which areas had reliable gusts of wind that would lift their skirts'. What is perhaps most significant here is that this practice was promoted by a boosterist press that was keen to argue that Montreal was modern, stylish and beautiful: arguing that it was a city particularly conducive to girl-watching helped support the idea that it was liberated, and feminized. For all this, changing attitudes towards the 'playboy' man about town, and the rise of a more visible women's liberation movement led to forms of resistance against such ways of looking at and categorizing women, with Wallace arguing that the streets of Montreal became more significant as sites for protest *against* the sexual objectification of women as the 1970s progressed.

Further reading: Wallace (2007)

> of the city . . . We adolescent girls, as a protective ploy, organized or categorized a whole range of what we called 'sex fiends'. There were a whole range of these sex fiends that made our lives a little more restricted our dreams a little darker. Subway fiends, roof fiends, park fiends and even museum fiends . . . and the dreaded movie fiend who groped his innocent young victim in the dark.
>
> (Nochlin 2006, 174)

This type of narrative stresses that women learn to perceive danger from strange men in public space despite the fact that statistics on rape and attack emphasize that they are more at risk at home and from men they know (women are up to six times as likely as men to experience domestic violence) (see Chapter Three on home spaces).

In this context, the rapist has been an urban figure whose persistent existence, both real and imagined, has served to place constraints on women's participation in the night-time city, whether this involves strolling to a take-away, walking home from a pub or nightclub, driving alone to an out of town cinema or the more mundane pleasures of walking a dog or jogging in a local park:

> The warning don't go into the park at night is more than just a simple notice of potential physical danger: it is also an acknowledgment of the shift in the park's function – which takes place when the sun goes down – from a place where nature lovers eat lunch and children feed squirrels to a place where one can score drugs or gets one's cock sucked.
>
> (Califia 1994, 57)

As such, it is more than simply a lack of light that encourages a feeling of vulnerability: it is the sexualization of night itself and the apparent giving over of the city at night to male pursuits. Feminist geographers have therefore identified fear of sexual attack as crucial in limiting women's access to, and control of, night-time space (Pain 1991). Surveys repeatedly note that fear of sexual and violent crime perpetuated by strangers in public space forces women to modify their behaviours or routes taken. For example, the 2008/09 British Crime Survey suggested that women fear rape more than any other crime, with 35 per cent of thirty-one- to sixty-year-old women reporting feeling very unsafe when walking in the city after dark, as opposed to less than one in fifty men. For some women this encourages a more profoundly domo-centric existence where staying in becomes viewed as infinitely preferable to dealing with the imagined dangers of the city at night (Whitzman 2007). For those who do go out, adaptations of behaviour can take different forms. Kavanaugh and Anderson (2009), for example, distinguish between individual-level risk management, whereby individuals modify their behaviour in specific leisure spaces, and environmental-level risk management, which can entail changes in the leisure spaces that an individual frequents. Examples of the former might include only visiting a venue in a group, carrying a personal attack alarm, avoiding interactions with strangers and leaving a venue in a taxi rather than on foot. The latter includes switching venues or avoiding specific spaces all together based on judgements of where risks are most apparent.

But the behavioural adaptations made in response to fear of sexual violence are more subtle than this and involve a combination of coping strategies that may be employed when in particular times and spaces, including a careful management of their 'embodied and spatialised selves' to avoid risks (Green and Singleton 2006, 862). One key tactic here has been the adoption of forms of dress that are seen as 'unsexy', noting that dressing 'provocatively' might be seen to attract unwanted or unwarranted attention (Solnit 2001). Indeed, histories of rape trials suggest that rape can often be transfigured through a not guilty verdict as an example of normal sexual behaviour, with forced sexual intercourse seemingly excused when women exhibit non-conforming behaviour by drinking alcohol, dressing provocatively or initiating intimacy. Such acts are deemed to send out signals of sexual interest that cannot be easily revoked when subsequently relied upon by an observer, especially in a context when male sexuality is commonly depicted as 'uncontrollably natural, consisting of overwhelming urges and desires, which leaves the male sexual imperative . . . unchecked' (Munro 2008, 923). In contrast, women are required to be strong, aggressive, and powerful in rejecting intimacy (Cowan 2007; Munro 2008). For Sedgwick (1993, 65), this illustrates the fundamental 'epistemological asymmetry of the laws that govern rape' and privileges the ignorance of men given it does not matter what the raped woman experienced or understood so long as the man raping her can prove he did not notice her lack of consent. Given such attitudes towards sexual attack, women's fear is generally regarded as 'normal' and their spatial assertiveness thought to be risky – something that can undermine attempts by developers to attract single women to live in inner city locations (see Kern 2007 and Case Study 5.3).

Sexing contemporary night-life: in what ways is the relationship between sex, gender and leisure changing?

This chapter has begun to identify some of the key sites of leisure which have opened up a diverse range of pleasures to the city's population. A key theme has been that these pleasures have been differently accessible to men and women. Many historical accounts of leisure therefore describe women's

CASE STUDY 5.3

'America's safest large city': policing rape in Philadelphia

As a distinct form of sexual violence, rape is a highly charged signifier of safety and danger as well as a legally ambiguous terrain. Given its narrow definition in the law (as non-consensual and unwanted penetrative sex), it is however a crime that is significantly under-reported, with victims either failing to see the event as 'rape' or lacking faith in the legal system to bring the perpetrator to justice. Moreover, many of the allegations of rape that are reported are invalidated by the police or are subsequently dismissed by the courts. Given that 90 per cent of rape victims are thought to be women attacked by men, this potentially means that a great deal of sexual violence perpetrated by men goes unreported and lies outside the ambit of legal action.

Considering rape in the US, Brownlow (2009) argues this under-representation of rape within crime statistics has been compounded by deliberate attempts to 'cook the books'. Contrasting rape with other crimes such as murders, Brownlow suggests that city leaders and police who are keen to demonstrate they are bringing crime rates down can easily 'hide' rape. He makes this point powerfully in his analysis of the rape and murder of Shannon Shieber, a twenty-three-year-old graduate student at the University of Pennsylvania. In the year before her murder, Brownlow details how four other rapes of single women living in the same area by the man (dubbed the City Center Rapist) ultimately identified as the killer went unpublicized by the police. In his article, Brownlow suggests that this was connected to a more systematic downgrading of violent crime in the city centre which fitted with attempts to brand it as a safe area in which to live and work. In particular, he dwells on the selling of the central city Rittenhouse district as a safe space for young women professionals at a time when Philadelphia was being touted as America's 'safest large city'.

In an era where urban safety has become a marketable commodity, and inner city housing has been vigorously promoted to young single professional women (see Chapter Four), Brownlow's analysis suggests that policing can be strongly influenced by urban policy. The fact that the rape of a young upper class woman in the city's most glamorous district was necessary to draw attention to the systematic downgrading and suppression of rape by Philadelphia police also points to a racialization of crime: for Brownlow, selling the city centre as safe for white women relied upon an unjust downgrading of the risk and sexual violence experienced by another population, namely low-income women of colour. Ultimately, he argues that women gentrifiers were attracted into the area by low crime rates that stemmed from years of police inaction and 'hiding' of sexual crime, and suggests that in the neoliberal city, the persistence of patriarchy and the machinations of property markets conspire to create the illusion of safety where there is none.

Further reading: Brownlow (2009)

---

engagement with spaces of leisure as being heavily circumscribed: for many women night-time leisure appears to have been important only up to the point at which they got married. The city at night itself could thus take on the appearance of a marriage market. Social histories of leisure suggest this role was particularly pronounced in the period of 'mandatory marriage' which emerged in the post-war period, with the late 1940s and 1950s having been described retrospectively as the 'golden age' of courtship because of the widespread assumption that people would enter into life-long heterosexual partnerships that would give their lives both meaning and structure. The subsequent growth in divorce rates and changing notions of intimacy and sex suggest that this was an exceptional period where particular ideas of modernity, security and romance combined to encourage the search for a life partner from a relatively early age (Langhamer 2007).

In pre-war Britain, for example, younger people had mainly gained introductions to the opposite sex through the 'evening walk', with many British cities having well-known routes along which groups of young women and men would walk, hoping to catch the eye of a potential date. Known colloquially as 'monkeyruns' or 'monkeywalks', these provided a relatively informal and regulated space for sexual performance and display, with well-known ambulatory circuits of the city being notorious for youthful groups and courting couples. However, with increasing affluence among young people in the post-war period, these public courtship rituals were superseded by forms of sociality played out in commercial leisure spaces such as dance halls, coffee bars, cinemas and cafés. Sometimes this was explicitly acknowledged: for example, many cinemas offered double 'love seats' for courting couples, with the cover of darkness of the cinema facilitating forms of sexual contact and intimacy that might have been discouraged elsewhere (Jancovich and Faire 2003). Films changed too, with more adult content becoming evident in the 1960s and 1970s. But other commercial leisure spaces also allowed for a sharing of intimacy, with the sensual pleasures of dancing, drinking and eating being important in courtship rituals, naturalizing the heterosexuality of desire and encouraging particular performances of masculinity and femininity. In this context, Meah *et al.* (2008) argue that spaces of evening leisure, where young women and men were allowed to mix freely, were particularly important sites for naturalizing the ideologies, identities and practices through which people entered heterosexual relationships.

The growing importance of commercial leisure spaces in the production of compulsory, coupled heterosexuality was thus something widely acknowledged in the mid twentieth century as evening leisure spaces became more available to younger, single and unchaperoned women and parental control waned. Since then, increasing autonomy of young people, the deferral of marriage and the rise of sexual permissiveness has seemingly led 'courting' and youthful heterosexual liaison to have spread to a much wider range of public spaces and sites, including youth clubs, sport spaces and unregulated open spaces (woods, parks, wastelands). For all this, formal leisure spaces remain important sites for 'coupling', albeit that the landscape of milk bars, cinemas and dance halls has been superseded by an arguably wider range of venue as the night-time economy becomes a driver of the post-industrial economy. Multiplex cinemas, bowling complexes, malls, arts centres, food halls, late night shopping centres and super-clubs all play a role as spaces for dating and the performance of coupledom.

In this context it is important to note that the revivification and promotion of the evening and night-time economy is currently a prime concern for many city governors given it is imbued with tremendous potential for creativity, sociality and profit-creation. Yet at the same time, the night-time economy has – at least in the urban West – become widely associated with forms of social disorder, noise, excessive alcohol consumption, violence and drug-related harm (Hobbs *et al.* 2007). As such, recent liquor licensing reform in the UK has sought to promote a more relaxed 'continental' ambience in the night-time city (Jayne *et al.* 2006; Roberts and Eldridge 2007). In theory, such changes are intended to make spaces of night-life more cosmopolitan, civil and open to a diversity of sexual performances and identity positions. Against this, the increasing corporatization of night-life, and its dominance by pub and club chains that promote high-volume alcohol consumption, continues to raise public and media concerns about violence and disorder, with many alternative and independent forms of night-life being squeezed out in favour of entertainment that seemingly panders to the lowest common denominator (Chatterton 2002).

What implications does the corporatization of night-life have for sexuality in the city? Given the 'static' of background violence that exists in the evening economy, and its apparent dominance by men fuelled by alcohol, it is tempting to suggest that it has normalized an essentially retrogressive attitude to sexuality by privileging predatory masculine sexualities. However, emerging ethnographic studies of night-life suggest that this is too much of a generalization, with leisure spaces remaining important sites where sexual and gender identities can be reworked. Clubs, for example, come in many different shapes and sizes, with Malbon (1999) showing that the sensual and embodied dimensions of dancing, the consumption of alcohol and/or drugs, the design and lighting of the club itself and the playing of particular types of music can combine to provide an ambience that encourages intense forms of socialization that can transcend existing distinctions of class, age, ethnicity or sexuality. Malbon's account implies that even in the most self-conscious and stylized settings, clubbing is about escapism and a suspension of the identity categories that characterize everyday life, with the experience of collectivity being a key appeal of clubbing for many participants.

Hutton (2004) also suggests that club spaces can be empowering for women, providing a comfortable space in which they can develop positive femininities based on a shared search for pleasure and excitement, sometimes allied with risk-taking behaviours. On the other hand, Hutton notes that not all clubs are the same, distinguishing between 'underground' spaces where it appears easier for women to feel comfortable and in control, and more mainstream clubs where she notes a tendency for men to view women as sex objects (or even 'meat' – hence the common description of large super-clubs in the UK as 'cattle markets'). Boyd (2010) makes a similar argument in the context of Vancouver's nightscape, contrasting 'Top 40' mainstream club venues with the alternative, indie clubs nights held in mixed venues in Vancouver's East End including warehouses, art galleries and down-market pubs. While she found the latter were characterized as safer and more comfortable by many of her respondents, the former were overwhelmingly described as sites of heterosexual aggression, violence and inebriation:

> I went to a mainstream bar one time called The Cellar. It's on Granville Street; you go downstairs. And I went there with some people I was working with on a movie, and I'd never been there before; it's like that whole Granville Street Strip, and I was appalled by what it actually was like. We just wanted to go there and dance. But we were dancing, and there was actually this semicircle of dudes around all of the girls, leering and then trying to get in and dance on you, and I don't know if it's normal in every club like that or just that club or whatever. I was appalled that that was fine, that that was allowed. I was freaked out by a guy and pushed him back, and he acted like I was overreacting! So there was like five of them; it was really, really, really, freaky . . . It was really scary.
>
> (Anonymous respondent, cited in Boyd 2010, 175)

Boyd argues that mainstream club scenes are dominated by a 'pickup' ethos, something emphasized in Grazian's work on the often hypermasculine performances played out in clubs and bars. Noting 'hot nightclubs and cool lounges enforce sexualized norms of dress and body adornment and invite flirtation, innuendo, and physical contact', Grazian (2008, 134) stresses that nightclubs are often settings where men work collectively through public displays of masculine bravado, aggression, humour and bodily display, often drawing their 'buddies' and friends in to support them in their sexual pursuit of women. Noting the interplay of camaraderie and competitiveness that is explicit in the ritual of the 'girl hunt', Grazian argues that the performance of masculinity and the search for social status among one's peers is more important than the search for sex itself in nightclub spaces. For all that, it appears that women in many mainstream night-life spaces can feel harassed by such rituals, and feel exposed to various forms of sexism (Chatterton and Hollands 2003; Kavanaugh and Anderson 2009).

That such forms of harassment remain prevalent within some mainstream spaces of night-life is remarkable in an era when young female consumers are increasingly important for the night-life industries and where many clubs have tried to move away from the 'cattle market' model (Chatterton and Hollands 2003). However, this is explicable in an industry which rarely employs women as DJs, managers, bouncers or promoters but often as dancers or barmaids. Many nightspots seemingly rely on the attractiveness of female service staff and the promise of eroticized interaction to recruit customers, with flyers, posters and adverts often featuring conventionally sexist images of women (see Figure 5.3) while male bouncers and door staff can play an active role in screening customers and embedding particular cultures of sexism, aggression and homophobia within nightclub settings (Hobbs *et al.* 2007).

Such hypermasculine cultures are perhaps most evident in the context of group tourism, and particularly during the homosociality of the 'stag' or 'bucks' party, which is inescapably part of the heteronormative institution of marriage (Thurnell-Read 2009). Neither truly spontaneous nor socially sanctioned, most stag nights represent a moment in which 'normal' masculine identities are played with, parodied and pushed to their limits at the moment when the 'stag' celebrates his 'last night of freedom'. With a traditional emphasis on the carnal body, stag parties often perform rituals in which the body is performed as an undecidable object of desire as well as an object of disgust: figuratively, and sometimes literally, the body is stripped, and takes on an indistinct position on the threshold between civility and nature. As well as marking the transition from single to married life, these rituals stand on the threshold between

Figure 5.3 Club flyer, Nottingham (source: Lauren Young).

what is deemed playful and what is regarded as profoundly anti-social, with the participants drinking competitively, 'girl-watching', acting lecherously and generally performing a hypermasculine identity. The celebration of a voracious, predatory male libido at stag events is often claimed to contribute to the general discomfort that many women feel when occupying male-dominated spaces of night-life (Measham 2004). As such, tourist authorities have sometimes voiced anxiety about stag tourism, suggesting that groups of men roaming the streets on stag celebrations stigmatize resorts and put off family consumers (e.g. 'Stag parties ruin Blackpool trade', *BBC News* 24 July 2003; 'Capital crackdown on stag and hen parties is on the cards', *The Scotsman* 25 March 2004). In Europe, the availability of low cost air travel has meant such anxieties have spread as certain destinations – Riga, Prague, Amsterdam, Kracow and Tallinn (see Case Study 5.4) – have become notorious as 'stag capitals' (Thurnell-Read 2009), while Mediterranean '18 to 30' resorts like Ibiza have long been represented as spaces where endemic drug and alcohol use fuels sexual promiscuity (in one survey, one-quarter of male visitors to the island reported having sex with more than one individual during their stay; half of the 75 per cent of male tourists who arrive on holiday without a sexual partner have sex while on holiday – see Bellis *et al.* 2004). As on the North American 'Spring Break', long associated with binge drinking, drug consumption and unsafe sexual practices, it appears that the liminal quality of being a tourist in an environment that is represented as permissive, encourages risky behaviour including excessive alcohol consumption, casual sex and irregular condom use (Maticka-Tyndale *et al.* 1998; Sonmez *et al.* 2006).

But while male homosociality in leisure space has been the source of widespread anxiety because of its association with aggressive male behaviour and predatory sexualities, the female counterpart of the stag do – the hen or bachelorette party – has also begun to attract attention (Tye and Powers 1998). Often, the hen party has also been read negatively as symbolic of a decline in female comportment, and taken as evidence for the emergence of a culture in which binge-drinking, sexual predation and excess are deemed acceptable for women as well as men (McRobbie 2008). This is often described by the media as an essentially lower class and uncivilized performance of femininity. Indeed, it has been argued that hen-partying women embody 'all the moral obsessions historically associated with the working class: a body beyond governance' (Skeggs 2005, 965). In Skeggs' own interviews with gay men frequenting Manchester's 'gay village', hens were variously framed in terms of contagion, pollution and distaste, their excessive heterosexuality making them literally *intolerable* for the gay clientele. In contrast to the male heterosexual stag party, the hen party was not seen to present the threat of violence, but was felt to undermine the comfort and ontological security associated with hard fought-for gay space. Skeggs contends that their heterosexual excess (e.g. dildos replacing necklaces) threatened to undermine the middle class respectability of gay leisure space (Skeggs 2005). In Casey's (2004) study of gay bars in Newcastle, it was also evident that women on hen dos gravitated towards gay bars, with the mixing and diluting of these venues viewed as problematic by the gay clientele (Casey 2004), who often questioned why 'straight' hen dos should even be admitted to gay venues given their tendency towards excessive behaviour.

Refuting such blanket characterizations of hen parties as disordered Eldridge and Roberts (2008) suggest that it is actually quite refreshing to see mixed age groups of women on the streets in towns and city centres where the dominant group remains (drunken) young men. They also argue that hen parties are not just about visits to male strip clubs and binge drinking, but also shopping trips, spa visits and extreme sports. While these are not always empowering or transgressive in themselves, Eldridge and Roberts (2008, 324) suggest that hen dos are too readily dismissed as constituting the 'wrong kind of feminism', and need to be viewed in the context of night-life that has been traditionally devoted to masculinist drinking cultures and where women's experience of public spaces at night have been strongly structured by fear. In this sense, hen dos can be seen as a symptom of a more assertive femininity – hence, the emergence of the 'ladette'. Day *et al.* (2004) argue that the description of women as ladettes is itself an attempt to exclude women from male drinking culture and space, a blatant attempt to reclaim alcohol consumption and related leisure spaces as male by suggesting that women behaving in the same way are acting in an inappropriate and essentially uncivilized way. The sexual dimensions of this characterization are clearly etched in many media descriptions, with newspaper and media exposés suggesting that the ladette's excessive alcohol consumption places her in situations where she might be vulnerable to the sexual advances of men (and 'date rape'), while she apparently gains social status among peers by talking openly about her sexual exploits. The popular television programme *Ladette to Lady*, for example, seeks to change young working class women's deportment, dress, manners and morality, suggesting that there is a clear line between being too sexual and not sexual enough in contemporary society, and that women's participation in night-life spaces needs to be informed by notions of middle class respectability and even chastity.

As this chapter has shown, such moral panics around women's participation in night-life are nothing new: in a historical analysis of 'troublesome' young women, Jackson and Tinkler (2007) suggest there are clear continuities between the way ladettes are represented in contemporary media and the

CASE STUDY 5.4

Homosociality, sex and excess: stag tourism in Tallinn

Tapping into anxieties about 'laddish' and 'yobbish' behaviour on Britain's streets, the contemporary stag night has become, for many, the prime illustration of the decline of *respectable* working class masculinity (notwithstanding the fact participants may be drawn from a wide diversity of class and cultural positions). This is something that is echoed in newspaper and media headlines about the behaviour of stag parties overseas, not least in Tallinn, Estonia. Since the break-up of the Soviet bloc, this walled medieval city has become a key focus of Estonian tourism, with ferry-based day trips from Sweden being superseded by low-cost airlines. Entry into the EU in May 2004 further raised awareness of the city as a tourist destination, increasing passenger arrivals from 868,000 passengers on scheduled routes in 2004 to 1.2 million in 2005 and boosting the tourist industry to one worth over one billion dollars annually (Jarvis and Kallas 2008).

Though Tallinn has been self-consciously branded as a historic tourist and as an 'eco-tourist' destination, it became evident that many of those visiting Tallinn in the 2000s with the low-cost carriers were on stag trips, many organized through tour operators specializing in Eastern European stag packages. One of these, Tallinn Pissup, boasts that the city is not just incredibly cheap but also has wild night-life: the company offers packages including 'lesbian strip shows', clay pigeon shooting, paintball, private limo tours and even a 'piss-up in a brewery'. Other companies specializing in Baltic stag tourism have offered similar packages focusing on combinations of daytime sporting activities and night-time leisure based on 'beer and babes', triggering rising concern about the reputation of the city:

> Tallinn no longer takes pride in the title of 'favourite destination of British staggers'. Recently, such visitors have become an increasing source of trouble. The Brits, loitering on the streets in large groups, swilling beer and making lewd suggestions to girls, are known all over continental Europe.
>
> (*Daily Telegraph* 7 August 2004, in Jarvis and Kallas 2008)

These types of exploit, recorded by a UK media keen to expose the misbehaviour of Brits abroad, encouraged some hoteliers and venues to put a ban on all-male groups, and, in 2008, the British Embassy in Tallinn went so far as to organize a workshop where they sought to promote responsible stag tourism, advising local stakeholders about British alcohol cultures and how to encourage more civilized drinking. At the same time, international media coverage began to identify stag tourism in the Baltic states as a driver of sex trafficking (see Chapter Seven) (Marttila 2008). Given this association with unruly sexualities, stag tourism remains more controversial than many other forms of alcotourism (Bell 2008), albeit the collapse of sterling against the Euro has somewhat stemmed the flow of UK men to the Baltic states on stag and group tourism.

Further reading: Jarvis and Kallas (2008)

---

descriptions of the flapper of the inter-war years. Their analysis thus highlights the ways hedonism, financial independence and sexual promiscuity have been attributed to both the flapper and the ladette. However, Jackson and Tinkler (2007, 267) argue that the most threatening aspect of the contemporary ladette, and that which causes most concern and panic, is her disruption of dominant discourses on gender and on women as carer. In their words, the 'ladette is presented as a pleasure seeker, with popular explanations for her "hedonistic tendencies" referring to women's increased financial independence and lack of family commitments'. The ladette is vilified precisely because she refuses coupledom, and seems to privilege sex over romantic love.

In this context, the contemporary city at night represents a space where women can escape *some* of the inequalities of the daytime city, and socialize with friends in a public manner that does not preclude discussion of sexual matters or assume coupledom is the only goal of women. The HBO series *Sex and the City* encapsulated this potentiality, suggesting that New York provided a multiplicity of spaces in which Carrie, Charlotte, Miranda and Samantha were able to enjoy the multiple pleasures of sex and shopping in each other's company (and sometimes that of gay men). While (heterosexual identified) men would often let them down, or provide unfulfilling sex, the city never did: moreover, it always appeared accessible and intimate rather than dangerous and unknowable, even at night (see Berman 2006; Handyside 2007).

All of this suggests that the varied forms of urban culture played out in the contemporary city at night provide opportunities for women – as well as men – to perform diverse subject positions and identities, and to 'play' with their sexuality as they engage with multiple spaces of consumption. Whether or not the types of evening leisure evident in the contemporary city actually respond to the desires of a broad base of the population is more doubtful given many spaces appear to encourage a very narrow range of heterosexual performances, and are often tied into forms of entertainment that privilege male spectatorship and sociability over female. And even though some women appear free to walk around the city at any time of night or day – as is the case for the post-feminist protagonists in *Sex and the City* – it has to be noted that this freedom is one bequeathed by racial and classed privilege. Leisure remains striated by the fault lines of desire, and the city at night should not be idealized as offering any sanctuary from the class, gender and racial inequalities that infest the daylight hours, no matter how heady and intoxicating its pleasures may appear at first glance. This is a theme that we return to in Chapter Six, which explores the rise of 'adult entertainment' as a distinct and important night-life sector, noting that despite progress towards the sexual liberation of women, the consumption of sexual entertainment by women remains relatively discrete and is barely discernible within the red light areas and 'hot zones' of Western cities that continue to cater predominantly for men.

Conclusion

Work on gay and lesbian identities suggests the city's night-life has gradually opened up to accommodate sexual difference in a variety of important ways (Hughes 2003; Slavin 2004). Despite this, this chapter has suggested that mainstream spaces of night-life remain distinctly heteronormative, being spaces where predatory and aggressive masculinities reign and where heterosexual rituals of flirting and coupling condemn women to take a more passive role. Within spaces of night-life, it appears women must appear sexy enough, but never be too sexy: while they are expected to perform powerful and (hetero)sexually desirable femininities, they are supposed to conform to middle class codes of respectability (Skeggs 2005). This tension has been highlighted through discussion of the 'ladette', a contemporary figure who promotes 'women's right to use public space, to be heard and seen, and to engage in pleasures that are considered relatively unproblematic for boys and men' (Jackson and Tinkler 2007, 270). From some perspectives, the sexually confident behaviour of women in the night-time economy is thus a victory for feminism, with the women effectively 'seizing the phallus' (Eldridge and Roberts 2008). Yet, for McRobbie (2008), and others, the phallic woman offers no critique of heteronormativity: it asks girls to perform behaviours associated with masculinity without relinquishing the femininity that makes them attractive to (heterosexually identified) men.

While the notion that the night-time economy remains fundamentally different to that of the daytime is problematic, this chapter has accordingly begun to draw out some of the key ideas concerning the sexing of night-time leisure, suggesting that this has tended to privilege the male gaze, and related masculine tendencies towards scopophilia, narcissism and voyeurism. The obvious corollary of this is that the nocturnal city effectively puts women's bodies on display so that they can be consumed by men. In effect, it appears that women are only allowed into the city at night on terms dictated by men, and must accept sexual harassment as a normal part of evening leisure. In contrast, there remains a popular stereotype that only men are able to handle the potential pleasures and pitfalls of the nocturnal city, viewing the 'dark corners' of the metropolis as 'regions of adventure, of challenging danger, or self-affirmation' (Schlör 1998, 171). However, such stereotypes have been strongly disputed in feminist critiques of the 'man-made city' (Rendell *et al.* 2000), and have triggered numerous responses, from 'Take Back The Night' marches and 'slutwalks' organized by women's groups (Nochlin 2006) through to the organization of women-only club nights and taxi services (Bromley *et al.* 2000). But rather than simply dwelling on the dangers of night-life for women, as well as for sexual minorities, this chapter has begun to identify the sexual pleasures that can be experienced by men and women in leisure spaces such as clubs, pubs, cinemas and restaurants (as well as the street spaces between them). Inevitably, however, some spaces of night-life are more obviously erotic than others: accordingly the next chapter turns to examine the consumption of sex itself, especially in emerging spaces of 'adult entertainment'.

Further reading

Chatterton and Hollands (2003) provides a wide-ranging, UK-centred overview of the production, consumption and regulation of spaces of contemporary night-life. Their chapter on sexuality touches on many issues raised in this chapter, considering the emergence of gay venues alongside a discussion of women's experience of more mainstream pubs and clubs.

Swanson (2007) provides a nuanced and fascinating historical account of the ways that new practices of consumption in post-war London were implicated in the making of new sexual orders, with the instabilities of sexual identity in spaces of cosmopolitan leisure and pleasure encouraging far-reaching debates about 'problem girls', prostitution and the limits of 'manly love'.

Turner (2003) provides a cultural reading of the city that focuses on practices of cruising, connecting the aesthetics of modern urban night-life to the pursuit of male pleasures in the cities of New York and London.

6 Consuming sex

Pornographies and adult entertainment

Learning objectives

- To understand the importance of the city as a marketplace for the selling of sexual images and materials.
- To become aware of the reasons for the centralization of sexual commerce within the economies of contemporary cities.
- To appreciate the anxieties that continue to surround the purchase and consumption of sex.

Historically, many key theories of urbanization and urban development have emphasized the city's role as a central marketplace for goods and services, noting the ability of cities to attract customers from a wide hinterland. As has been emphasized in previous chapters, sex is one commodity that is bought and sold in cities – one that is often highly valued – yet its importance in supporting the growth and expansion of cities is seldom acknowledged in this literature. In part, the reticence among urban theorists to acknowledge that sex commerce makes an important contribution to urban economies is related to the fact that sex markets often exist outside, or alongside, the formal economy (Sanders 2009b). As was described in Chapter Two, the sale of sex often occurs out of sight of the state and law, behind closed doors and 'in the shadows'. As such, the contribution that sex businesses make to the urban economy is often hard to gauge.

Despite the evidential problems faced by researchers seeking to quantify the sale and consumption of sex in the city, this chapter demonstrates that there is a growing, if sometimes grudging, recognition that sex businesses are part and parcel of urban economies, and can make important contributions to job creation and wealth generation. In part, this recognition has resulted from city governors themselves beginning to argue that 'sex sells', suggesting that sex businesses can be important in creating vibrant and vital city economies. As Chapter Five described, this is particularly the case in post-industrial cities, where reliance on manufacturing has given way to service- and leisure-based economies in which the evening economy of pleasure, leisure and entertainment is particularly significant (Chatterton and Hollands 2003). Tellingly, within this playful, leisured *experience economy*, sex businesses are being redubbed as 'adult entertainment', a euphemism enthusiastically adopted by a sex industry keen to be accepted as legitimate and respectable (Comella 2010).

This chapter hence describes two related tendencies in the contemporary city: on the one hand, the *sexualization* of mainstream sites of leisure and consumption and, on the other, the *corporatization* and up-scaling of sex businesses that were once themselves on the margins. It considers the extent to which these tendencies have normalized sex businesses within the urban economy, noting that for every voice arguing that sex businesses are now respectable, stylish and upmarket there appears to be a contrary voice condemning such businesses as immoral, exploitative or demeaning. It does this by focusing on some of the spaces of sexual consumption that have become increasingly central and visible in Western cities, particularly the sex shop and the lap-dancing (striptease) club. In this sense it is extremely useful to begin by considering accounts that identify the increasing prominence of sex businesses in the city as one symptom of a wider *pornification* of society (Smith 2010).

Mainstreaming sex commerce: in what ways is society becoming pornified?

As Chapter Two showed, the purchase of sexual services has often been regarded as socially problematic, depicted as a threat to urban moral order. Yet, on the other hand, some have countered by arguing that prostitution fulfils a 'useful' social function, providing an outlet for the city's 'excess' sexual energies and allowing people to experience forms of sexuality that they might otherwise be denied within their long-term (assumed monogamous) relationships. Traditionally, these arguments have also been widespread in relation to the consumption of pornography, a commodity form indelibly associated with urban life. Though sexual images had long been included in satire, philosophy and medical texts, it was only in the late eighteenth century that representations apparently began to emerge that had the sole aim of sexually arousing the viewer (Hunt 1996; Kendrick 1996). Certain key technologies (e.g. printing, photography) and sites (e.g. engravers, bookstores, newsagents) were crucial in circulating these representations (McCalman 1988), with the threat of mass reproduction leading the police, judges, readers and librarians to talk of a particular set of images and representations as 'obscene' and in need of regulation. As such, the city itself was implicated in the production of a new cultural form that might be appreciated and enjoyed by the 'educated' (male) consumer, but which was deemed to pose a threat to the innocent who would potentially be seduced and corrupted by exposure to its sexualized forms (Sigel 2002). The free circulation of obscene materials in the city was a challenge to the bourgeois male mastery of the streets in much the same way as the visible presence of the female prostitute in the city's thoroughfares was: from the 1850s there was talk of a pornography 'epidemic', and numerous attempts to limit access to materials that might 'corrupt and deprave'.

The cultural historian Lynda Nead (2005) offers a particularly insightful account of the emergence of pornography as an object for regulation in the context of bourgeois fears and fantasies about the modernizing metropolis. Arguing that 'obscenity was the brash new produce of Victorian London' (Nead 2005, 151), her examination of the anxieties surrounding the display of obscene prints and lascivious images considers the ways in which these were contrary to the economic, social, aesthetic and moral aspirations of those who were seeking to make the city more commodious. Suggesting that such images were an 'inevitable by-product of the new rituals of urban leisure and commerce that accompanied Modern urbanism', Nead (2005, 148) hence considers the rhetoric of those who felt that the torrent of explicit and licentious images bombarding 'unsuspecting pedestrians' from all sides threatened to draw 'respectable' citizens into a state of sexual fantasy and depravity (Nead 2005, 149). Her examination offers Holywell Street as a microcosm of these debates, a small side street off the Strand in London that was synonymous with obscenity in the mid nineteenth century because of its association with the printing and display of licentious cartoons, smutty guidebooks and 'French specialities'. For this reason, Holywell Street became a site of particular scrutiny in the context of urban improvement. Described as a 'spatial aneurism', a labyrinthine and dark thoroughfare that stood in contrast to the bright, wide and straight thoroughfares of respectable London, the language of the civic improvers emphasized the dangers associated with the dark side of desire, metaphorically linking death and the erotic (see Bataille 1970). But Holywell Street was not depicted simply as a local difficulty – a concentrated site of spatial and moral confusion – but a major source of London's moral impurity. Significantly, the threat the street posed was linked to its Jewish past. As Nead (2005, 176) explains, 'the Jewish traders provided the mythic dimension of the place; they were ciphers for the dangerous transactions that were imagined in the dark confines of the narrow lane'. In this way, Holywell Street became symbolic of the Other, both racial and sexual. Lord Campbell, the author of the first UK Obscene Publications Act, hence looked forward to a time when Holywell Street would be 'the abode of honest, industrious handicraftsmen and a thoroughfare through which any modest woman might pass' (cited in Hunter *et al.* 1993, 63) and, following the introduction of the Act in 1857, it was no coincidence that the first person arrested was the pornographic publisher William Dugdale, who sold and distributed prints from 51 Holywell Street.

Nead's discussion of the 'threat' of obscenity in mid Victorian London usefully highlights some of the gendered double standards that have been associated with debates around the sale and display of pornographic goods. In particular, it stresses that men have often argued against the free availability of pornography not necessarily because they do not wish to see it – and it is clear that in some cases legislators are keen consumers of pornography – but because they are concerned that more susceptible groups might be corrupted by it. In the debates surrounding the passing of the Obscene Publications Act 1857, pornography was deemed infective and its female 'victims' portrayed as being tempted to follow a path of sexual transgression that could feasibly culminate in death (Hunter *et al.* 1993). Subsequent Acts made similar attempts to justify control and classification of the obscene based on similarly patriarchal and paternal notions, in some cases enshrining ideas that some materials should remain accessible to 'incorruptible intellectuals' for aesthetic and scientific reasons, yet should be strictly off limits to 'wives and servants' (see Hornsey 2010 on the UK's 1959 Obscene Publications Acts and the 1960 trial concerning D. H. Lawrence's infamous book *Lady Chatterly's Lover*). Yet in subsequent decades, the collapse of established cultural categories in the wake of 1960s countercultural movements (Collins 1999), coupled with the rise of new technologies of transmission (especially the video cassette, and later DVDs), witnessed important shifts in the accessibility of pornography, with it becoming more available, but conversely, more contested. In the 1970s and 1980s in particular, high visibility campaigns inspired by the work of Andrea Dworkin and Catharine MacKinnon, among others, began to argue that sexual content should not be suppressed because it violates community standards of decency, or threatens to draw women into immorality, but rather because it represents a form of gendered exploitation. In the work of MacKinnon (1993), for example, porn's preoccupation with the penetration of (submissive) women's bodies by men was regarded as a form of gendered violence in and of itself, described in visceral terms as a political practice subordinating women, legitimizing rape and encouraging sexual exploitation. In numerous other accounts, pornography is accused of being a key site of women's sexual objectification, with the content of mass marketed sexual representations effectively reducing women to a mere commodity to be bought and sold, stripped of agency and depicted as a body-in-parts (Jensen and Dines 1998; Jeffreys 2008).

While anti-pornography feminists remain a vocal lobby in debates around commercial sex, often working alongside religious and neo-conservative groups arguing for the protection of 'family values', the argument that pornography remains a bastion of gendered exploitation has been recently challenged as new understandings of sexuality have come to the fore. These include the idea that the consumption of pornography can be both normal and healthy, and that there is no necessary contradiction in 'paying for pleasure' (Williams 1989; Paasonen 2009). Associated very much with the transformation from modern, industrial societies to late-modern, knowledge-based societies, key here is the idea that sex can be recreational, and not merely procreational (see also Chapter Two on sex work). Giddens' (1992) notion of *plastic sexuality* exemplifies this argument given it emphasizes the malleable nature of sexuality, which can be moulded according to individual desires. For a commentator like Giddens, the recognition of sex as plastic marks the possibility of an emancipatory sexual politics in which binary distinctions of good and bad sex (i.e. between marital/procreative and non-marital/non-procreative) collapse, to be replaced by a sexual morality based on 'pure relationships': sex for its own sake 'continued only in so far as it is thought by both parties to deliver enough satisfaction for each individual to stay within it' (Giddens 1992, 58). The transformation of intimacy thus allows for the forging of more equal relationships between men and women, with pornography being one site where, theoretically, both can find pleasure and fulfilment free from the traditional constraints of patriarchy. Such arguments resonate with the rhetoric of sex-positive feminists, who, contra the arguments of Dworkin and MacKinnon, claim in different ways that it is important that women have access to diverse sexual images and performances, and that attacks on the sex industry ignore the positive experiences that many women can have as producers or consumers of adult entertainment (see, for example, Williams 1989; Kipnis 1996; Smith 2007b; Paasonen 2009). Indeed, despite tendencies towards corporatization and global standardization (see Chapter Seven), it appears that the sex industry promotes a diversity of sexual representations and fantasies that do not just promote male sexual dominance and female subordination (Smith 2007a). Porn has hence been figured as a site where women and sexual minorities can critique heteronormal values through an emphasis on the 'authentic', empowered body (Langman 2008).

Figure 6.1 Men's interest only? Pornographic magazines on the British high street, 2008 (source: Y23/Creative Commons).

The rise of sex-positive discourses and queer readings of pornography are among the factors leading to what Brian McNair (2002, 137) famously described as the pornification of society: a mainstreaming of sexual images and iconography into popular media forms, to the extent that the codes and conventions of pornography are 'part of the armoury' of popular cultural production. In other accounts this is presented in terms of the rise of 'raunch' culture and 'porno chic', a sexualization of culture that has seen sex becoming increasingly visible – and increasingly explicit – within the mainstream (Attwood 2005). An example of this is provided by the advertising campaigns of American Apparel, now the largest producer of clothing within the US. Founded in Montreal in 1989 by Dov Charney, the chain promotes itself as an ethical retailer that promotes workers' labour rights and is opposed to exploitative practices reliant on 'sweatshop labour'. The company has also made much of its support for gay and lesbian rights, campaigning against the Californian repeal of civic partnership in 2008. This appeal to the sensibilities of young, ethically reflexive consumers is something that the company seeks to promote through its advertising campaigns (designed in-house), which avoid airbrushed images of supermodels in favour of grungy, edgy pictures that emphasize values of authenticity. Yet the images are also predominantly of women, and young women at that, often in poses that are sexually suggestive (rather than explicit). Eschewing the normal rules of advertising, and using models that are often employees, the company's images can easily be mistaken for soft-core 'amateur' porn. This has often courted controversy, and one set of images that appeared on the back cover of a high street fashion magazine (that featured a women wearing an American Apparel hooded top and underpants gradually exposing more flesh, and ultimately a nipple) was banned by the UK Advertising Standards Authority (ASA) because it appeared to sexualize a model who appeared to be under sixteen (albeit she was in fact twenty-three). In its judgment, made in September 2009, the ASA concluded that the advert should be banned for reasons of taste and decency, and stated that the photographs appeared as if the model 'were stripping off for an amateur style photo shoot'. In this sense, the complaints about this advert were themselves predicated on knowledge of the cultural and aesthetic conventions of porn, underlining that even when its presence is contested, pornography is already in the mainstream.

The idea that a high street retailer with stores in more than twenty countries, and over 10,000 employees worldwide, has chosen to adopt an explicitly sexualized market strategy at the same time as making major claims to corporate social responsibility ostensibly symbolizes a shift in the moral landscape, and a sexualization of mainstream retailing that has been much commented on. Other examples of this putative mainstreaming of sex are legion, whether it is the proliferation of soft-core 'lad's mags' across the shelves of British supermarkets; the availability of sex toys, vibrators, lube and cock rings in high street chemists that once drew the line at condoms; thousands of DVDs and books advising people how to 'spice up' their love life; the proliferation of unscrambled free-view television channels like *Babestation*, *House of Fun* and *Babeland* which normalize the consumption of soft core porn; or major supermarkets selling pole-dancing and striptease kits (Attwood 2005; Marriott 2009; Dines 2010). Often, writers listing such symptoms of pornification suggest this is having erosive effects on society, with the prevalence of sexual content online and in public spaces suggested to be disturbing children's 'natural' sexual development; causing sex addictions that lead to marriage breakdowns and encouraging people to mimic the hypersexualized lives of porn stars without an understanding of the health risks involved. However, are there benefits of living in 'sexed up' societies? Is the increased availability of sexual content allowing for a fuller and richer discussion of possible sexual lives, and educating people about what constitutes a healthy or good sex life?

The answers to these questions are a matter of intense debate. However, many commentators suggest that the proliferation of sexual content across multiple media needs to be viewed alongside an acknowledgement of the diversification of that material, and an escape from the narrower range of possible sexual scripts offered in conventional 'plotted' heteronormative pornography. There is now pornography seemingly catering to all sexual tastes, including, among other genres, gay porn, BDSM and fetish porn, 'inter-racial' porn, amateur porn, porn filmed in public places, reality porn, mature porn, point-of-view 'gonzo' porn, voyeur porn and alt.porn, a largely independent genre that is allied to sub-cultural movements like goth, and often features performers who are pierced or tattooed (Jacobs 2008; Langman 2008).

For all this diversity, most commentators writing on pornography and the sex industries argue that mainstream culture and the adult commercial sex industries are, in many important senses, converging (McNair 2002; Bernstein 2007; Brents and Hausbeck 2010). For Brents and Sanders (2010), this mainstreaming needs to be considered as having two distinct dimensions, namely, economic and social:

> Economic mainstreaming can involve changes in business forms, marketing, and distribution whereby sex businesses look and act like majority, conventional, ordinary, normal businesses. For example, mainstreaming sex-industry businesses can adopt traditional business forms such as corporate structures, vertical and horizontal integration, chains, franchises, marketing techniques, and traditional forms of financing . . . Social mainstreaming shifts cultural attitudes toward the acceptability of sexuality as a legitimate form of commerce and pushes businesses toward smoother integration with mainstream social institutions. Bodies, physicality, and sexuality as modes of commercialization in all aspects of consumerism have allowed the direct and indirect purchase of sexual services to become more visible and accessible on the high street and in public spaces.
>
> (Brents and Sanders 2010, 57)

As this quotation suggests, the most obvious urban outcome of such mainstreaming is a shift of sex businesses from back alleys, 'beneath the arches' (Kolvin 2010) and 'municipal districts of ill repute' (Liepe Levinson 2002, 22) to the heart of the city's consumer landscape. Ryder (2009) argues that traditionally, sex businesses have been peripheral to the social and economic life of the city, characterized by transience and a generally run-down air. More recently, it appears that they have opened up in more mainstream and respectable spaces, in prime shopping streets and even in retail malls (Hubbard *et al.* 2008).

Brents and Sanders (2010) illustrate this with reference to two contrasting cities: Las Vegas and Leeds. While the former is a major tourist destination with an international reputation, the latter is a provincial centre that has little tradition of tourism and was, until the 1970s, a centre of engineering and manufacturing. Yet in both they suggest that the increased reliance on services and the entertainment economy has encouraged the centralization of businesses offering indirect sexual services, mainly in the form of lap-dancing venues (Leeds has twelve such venues, Las Vegas many more). In both cities they suggest that the consumption of sex has become highly visibilized, accessible and normalized, tacitly encouraged by urban governors who regard these as an integral part of the night-time leisure economy. In some other cities, this encouragement is even more obvious, with politicians having publicly promoted striptease and adult entertainment by arguing for liberalization of the laws restricting the opening of such businesses (see Case Study 6.1).

The notion that sex businesses have been centralized both socially and spatially must, however, be tempered with the observation that cities like Portland, Oregon are exceptional in placing few restrictions on the opening of such premises. In the majority, there remain a host of legal constraints and social norms that seek to restrict, repress or reduce the visibility of sexual content, meaning sites of adult entertainment, be they sex shops, sex cinemas or strip clubs, often constitute fiercely embattled sites, facing campaigns of opposition that can ally anti-pornography feminists, religious groups and neo-conservative pro-family advocates. The success or otherwise of such campaigns is of course place-dependent, and whether a sex-related business is permitted to open in a given circumstance may be dependent on the decisions made by those who have jurisdiction within a given territory, who are required to consider the views of 'experts' such as planners, licensing officers, solicitors and the judiciary as well as those who oppose such businesses on the basis of their immoral or obscene nature (Hadfield and Measham 2009; Hubbard 2009).

The extent to which sex businesses are allowed to flourish in city centres accordingly varies considerably from city to city, dependant on their perceived suitability in different communities. For instance, planners may regard 'adult cabarets' as acceptable in a particular location, but not an X-rated bookstore. The suggestion here is that the acceptability of sex-related businesses in particular locales is adjudged by regulators who consider the nature of both the business and its location: examination of these contested knowledges is hence vital if we are to understand why commercial sex flourishes in some spaces, but not others. To explore this further, the remainder of this chapter will examine some of the ways the state and the law seek to

CASE STUDY 6.1

Striptopia in the US: adult entertainment in Portland, Oregon

With a population of 600,000 and around fifty dedicated striptease and 'exotic dance' venues, Portland, Oregon can claim to have perhaps the highest number of adult-orientated businesses per head of population of any city in the world. By dint of that fact, adult entertainment has become an important part of the city's civic identity. Moreover, given these clubs are distributed widely across the city, and not clustered in any sort of 'red light zone', it is hard to avoid drawing the conclusion that striptease is an integral part of the city's economy – and a key facet of its self-representation as a vibrant, quirky and multicultural city that is 'green' and friendly: significantly, Portland boasts what is thought to be the world's first vegan striptease club, Casa Diablo, and a varied range of clubs catering to the city's diverse sexual communities. The city is widely acknowledged to be a centre for 'new' burlesque dancing, which reinvents iconic 1940s traditions of striptease and bawdy cabaret in ways that can be seen to be transgressive and even queer (Ferreday 2008). There are, in effect, forms of sexual entertainment catering for those with a wide variety of tastes, including fetish and BDSM themed clubs as well as those catering to women-only and non-white ethnic minority audiences.

The visibility and prevalence of strip clubs in Portland was initially facilitated by a 1982 state ruling which decided that all communication deserved protection as free speech (a decision that contrasts with US federal judgments), with subsequent test cases suggesting that this extended to include 'obscene' speech. Although there have been subsequent ballots designed to challenge this permissive interpretation of free speech and to 'clamp down' on adult entertainment, none have succeeded. This means that the city has no right to treat sex businesses differently to any other business, and also that it has no right to censor strip club performances on the basis of performers' attire (total nudity being permitted). The city's first all-nude 'gentleman's club' opened in 1989, permitted on the basis that nudity was allowed in other city spaces, including a beach, in art classes and even at a country fair. As the popularity of such spaces became clear, smaller bars and pool taverns began to add dance stages to compete with the larger strip club operations (McGrath 2010).

While McGrath relates the growth of Portland's striptease economy to the decline of its traditional fishing and timber industries, Sanchez (2004) goes somewhat further in relating this to the imperatives of neoliberal urbanism, arguing that the normalization of adult entertainment in the city is related to the efforts made by city governors to market the city as a vibrant night-life destination at the same time that they have sought to exclude abject forms of sex labour from the city. Contrasting the laissez-faire policies of the city governors in relation to striptease clubs with the more repressive 'anti-prostitute' zoning ordinances enacted in the city, she points out that even if the performance of striptease has become normalized in the city's leisure economy, the performers themselves remain excluded from full workers' rights, being classified as 'private contractors' who have to pay for the 'right and privilege' to participate in sexual commerce, and live off the tips left by customers (Sanchez 2004, 881). This suggests that the social and economic mainstreaming of adult entertainment is not always a process that proceeds in parallel with improvement in workers' pay and conditions (see also Hubbard *et al.* 2008).

Further reading: Sanchez (2004); McGrath (2010)

---

influence the location of adult entertainment – and in some cases prevent the de facto pornification of urban space.

Planning for sex businesses: how does the state influence the location of adult entertainment?

Although obscenity and censorship laws can be used to control the type of goods or performances provided by sex businesses, changing national standards for obscenity mean that local government has often felt it necessary to introduce 'command-and-control' techniques, including licensing, zoning and planning powers to organize and even micro-manage spaces of sex consumption (Ryder 2004). In the US, for example, zoning ordinances have been widely used since the 1970s to prevent adult businesses opening in particular locales, typically prohibiting sex-related land uses near homes, schools and religious facilities, effectively restricting businesses to industrial districts (Kelly and Cooper 2000). Though such attempts at control have been contested, not least by those who argue that zoning restricts rights to free expression (Liepe Levinson 2002; Hanna 2005), the US Supreme Court confirmed the legality of adult use zoning restrictions in 1976 when Detroit's 'Anti-Skid Row Ordinance' was upheld in *Young v. American Mini Theaters Inc.* on the basis it served a 'substantial governmental interest' (namely, reducing the 'negative secondary effects' surrounding adult businesses) (Edwards 2009). Municipalities wishing to prohibit adult businesses from operating in certain areas have thus cited a need to protect communities from alleged effects such as increases in crime, decreased property values and neighbourhood deterioration, drawing on a number of (possibly flawed) studies from the 1970s which indicated that, compared with other land uses, there are increased crime rates and lower property values near sex-related businesses (Linz *et al.* 2004). On this basis, federal courts have accepted that high concentrations of adult businesses damage 'the value and integrity of a neighbourhood', further stipulating a city 'does not need to wait for deterioration to occur before setting out to remedy it' (*15192 Thirteen Mile Road v. City of Warren*, cited in Tucker 2007, 420).

Alongside zoning, however, is licensing. Governmental licensing involves the stipulation of particular controls over a defined area of activity, and usually relies on systems of monitoring, inspection and policing to ensure compliance with rules. The consequences of non-compliance usually involve licence revocation, fines or, rarely, imprisonment. For example, the forms of licensing evident in the US vary from state to state, with the type of control that can be exercised over sexually orientated businesses dependant on whether the business serves alcohol or not. Licences may accordingly be granted subject to certain conditions which, in the case of a sex business, can relate to the character of the owner, the operating hours of the establishment, the nature of the goods or performances sold as well as the security measures enacted on the premises (Hanna 2005). Though zoning conditions may seek to control some of these operational issues, once a zoning has been granted there is little discretion to adapt the zoning or impose more stringent conditions. Licences, in contrast, are generally renewed annually, and may be revoked for failure to uphold the conditions (Kelly and Cooper 2000). Licencing provides a flexible means by which the state can ensure that those who make their living selling 'risky' pleasures take responsibility for managing and running their business in accordance with the stipulations of the local state. Licensing also absolves the state from subjecting individual premises to constant surveillance, as secondary evidence of licensing infractions can be given material weight in any application for a licence renewal (Hubbard *et al.* 2008).

In contrast to zoning, licensing has a more lengthy history in the US and elsewhere as a means of regulating spaces where sex is on display. Friedman (2000), for example, reviews the decade-long campaign against burlesque entertainment waged in New York by religious, anti-vice and municipal activists in the 1930s, led by mayor Fiorello LaGuardia who accused this entertainment of propelling working men to seek adulterous liaisons, abandon their families and jeopardize their workplace productivity (Ross and Greenwell 2005). Initially obscenity laws were used to censor the content of burlesque, yet the fact that striptease artistes wore flesh-coloured underwear meant that it was difficult for opponents to substantiate allegations of explicit sexuality or nudity. As such, Property Owner Associations, concerned burlesque shows were lowering the tone of particular neighbourhoods, pressurized New York's licensing commissioner not to renew burlesque venues' licences on the basis of their clientele – and not the performance per se – alleging that threatening crowds of men would congregate on the streets outside theatres, harassing female passers-by. Friedman hence argues the eradication of burlesque in New York was underpinned by anxieties about the disorderliness of the male working class audience: by 1942, every burlesque theatre licence in New York had been revoked on the grounds they were the habitats of 'sex-crazed perverts' (Friedman 2000, 87).

While zoning and licensing can theoretically combine to repress sex businesses and 'plan them out of existence', the more common strategy has been one of containment. Collectively, both zoning and licensing laws can thus be identified as complicit in the creation of 'red light districts' (see also Chapter Two on the policing of sex work). Ryder (2009) argues that creating such districts can bring benefits for customers (who are able to easily compare prices and goods because of the clustering of premises) as well as non-customers (who can avoid such areas if they are offended by the sight and sounds of the sex industry). Moreover, he contends that these areas are ones of potential economic growth, being located in relatively central but underdeveloped neighbourhoods that might be described as 'incubation districts' where sex businesses often co-exist with other marginalized land uses and cheap rental properties. In many instances, red light areas have become associated with creativity in sectors including art, music, fashion and film production, being described as 'neo-bohemian' spaces characterized by individuality, openness and diversity (Fougere and Solitander 2010). An example is the Reeperbahn, in Hamburg's St Pauli district, which over the years has been associated not just with sex but also art and music (and most famously, The Beatles, who set up residency there in the early 1960s). Red light districts can thus attract a mix of business start-ups, meaning these areas can be viewed as competitive urban assets rather than liabilities (Ryder 2009).

While restricting commercial sex to well-known red light districts via licensing and zoning powers thus has some advantages, experience suggests they can conversely become something of an embarrassment to city governors if they become known as an area of 'sleaze' at odds with the image that the city wants to project. A much discussed example here is Boston, Massachusetts, which used zoning powers in the 1970s to bring commercial sex together in a well-known 'combat zone' between Boylston and Kneeland

Street. Rising rates of crime in the area, and the highly publicized murder of Harvard footballer Andrew Puopolo in 1976, led to a rethink, with the reversal of zoning ordinances and an attempt to disperse sex businesses to the city fringes (Tucker 2007; Phan *et al.* 2010). Similar strategies were pursued in the Times Square area of New York in the 1980s, which by that time had become a 'no-go' area for many women (and some men) because of the concentration of (poorly managed) sex cinemas and 'girly' shows:

> The antithesis of productive social space, Times Square . . . was shunned by developers, the middle classes and mainstream retailers alike as a kind of no-man's land saturated, in the popular imagination, with sex, sleaze and criminality.
>
> (Papayanis 2000, 351)

At city mayor Rudolph Giuliani's request, New York City Council hence approved amendments to their Zoning Resolutions in October 1995 designed to 'encourage the development of desirable residential, commercial and manufacturing areas with appropriate groupings of compatible and related uses and thus to promote and to protect public health, safety and general welfare' (cited in *New York Times*, 23 February 1998). In effect, this resolution forced the closure of non-compatible land uses, including adult establishments. The new law characterized 'adult establishments' as 'objectionable non-conforming uses which are detrimental to the character of the districts in which (they) are located'. Dictating they should 'be located at least 500 feet from a church, a school (or) a Residence District', and 1000 feet from another adult entertainment, this law represented a remarkable attempt to reaffirm socio-spatial order by maintaining clear distance between pornography and 'family' spaces (Papayanis 2000).

Despite mixed evidence that the presence of adult businesses actually impacted negatively on property values in the Times Square district (Insight Associates 1994), or promoted significant criminal activity, its subsequent reinvention as the 42nd Street Precinct Business Improvement District is often cited as a key milestone in New York's 'battle' against X-rated businesses (Boyer 2001; Liepe Levinson 2002; Miller 2002; Riechl 2002). In turn, this paved the way for the area's reinvention as a 'family-friendly' entertainment district, with the Disney Corporation being a key investor in the area. On this basis, such 'anti-porn' ordinances have since become widespread in the US, with the 'success' of New York in cleaning up its sex businesses mirrored in similar efforts to 'clean up' areas of sex business in other countries (see Case Study 6.2 on Amsterdam). Yet it is a moot point whether planning powers of zoning and licensing respond to community or public anxieties about

CASE STUDY 6.2

Zoning out the sex industry: gentrifying Amsterdam's red light district

Ashworth *et al.* (1988) identified Amsterdam – alongside Berlin – as Western Europe's 'sex capital' because of the persistence of visible prostitution at the heart of the tourist city. This is partly the legacy of the 1911 law banning brothels in the Netherlands which encouraged women to set up businesses independently, advertising their services by sitting behind their windows. Latterly, entrepreneurs and club owners established 'window working' on a more commercial basis, with the presence of prostitution encouraging a diverse range of sex shops and clubs within a distinctly touristic 'red light area' along the Zeedijk (the old sea wall) and in the adjoining streets of the Wallen.

Describing this area, Wonders and Michalowski (2001, 553) suggest it resembles a 'modern open-air shopping mall' of 'sex clubs, sex shows, lingerie and S&M clothing shops, condomeries, and a sprinkling of porno shops' in a clean, safe and historic setting. However, the notion that the area is clean, commodified and safe for tourists has not always been dominant, with Ashworth *et al.* (1988) noting the personal insecurity that was associated with it prior to the police 'sweeps' that gradually removed the drug-dealers whose consumer base largely consisted of addicted street working prostitutes and unwary tourists (Verbraeck 1990). Given such problems of drug-dealing and anti-sociality persisted in these areas, the City Council established a toleration zone as the most effective way of managing street sex working. As Visser (1998) describes, the 'tippelzone' (literally, strolling zone) was established in 1996 in the docks area behind the Central Station partly to 'clean up' the historic Wallen area. From 2000, legalization of the sex industry allowed both brothel owners and those running sex clubs to buy licences: unlicensed clubs and those outside areas designated for sex consumption have been closed down.

Despite this, the situation in the Wallen remained of concern to the authorities, with increased stag tourism prompting complaints about 'shouting in the streets, clients who urinated in people's doorways, a generally threatening atmosphere, much criminal activity of drug addicts and pimps, and frequent muggings' (Wagenaar 2006, 14). This reputation for (alcohol-fuelled) rowdiness in the Wallen became seen as a major obstacle to marketing Amsterdam as an attractive destination to other holidaymakers, including gay tourists (Kavaratzis and Ashworth 2007, 18). Through judicious use of licensing laws, the Amsterdam Centrum Council thus begun a process of 'thinning out' existing sex business, using the 2006 'Bibob' law which allows the authorities to use forensic accountancy procedures to identify where premises managers might have been involved in illegal activities in the past. This has allowed the Amsterdam Centrum authority to revoke the licences of half of those businesses initially granted a licence in 2000, leading the international media to speculate that Amsterdam's red light district is being effectively closed down (e.g. 'Amsterdam closes a window on its red light trade' the *Observer* 23 September 2007; 'Amsterdam to curb red light district' *Houston Chronicle* 14 October 2007). In addition, Mayor Job Cohen oversaw the compulsory purchase of eighteen further premises.

Nonetheless, as of 2007, the Wallen still boasted fifty-seven sex video shops and sixteen sex clubs and theatres. Rather than being an attempt to 'close' the red light district, the withdrawal of licences for some premises – and the compulsory purchase of others – can be seen as a response to this demand for more upscale tourism, meaning that sex businesses in the Wallen now co-exist with designer clothes shops and restaurants – and not just other sex premises (see 'Amsterdam tries upscale fix for red light district crime', *New York Times* 24 February 2008).

Further reading: Wonders and Michalowski (2001); Nelen and Huisman (2008)

---

commercial sex or whether they serve the interests of property developers seeking to profit from areas 'blighted' by sex businesses: the zoning laws adopted in New York, for example, have been widely interpreted as working in the interests of corporate developers (Papayanis 2000), while Ross (2010) contends that the 'cleansing' of Vancouver's red light district was a thinly veiled attempt to make it safe for 'bourgeois capitalism'.

In accordance with the *revanchist* city politics that seeks to remove potential threats to property-fuelled gentrification (Smith 1996), it is possible to suggest some sex-related businesses are viewed as obstacles to the cultivation of a leisured and profitable 'glamour zone' at the heart of Western cities. Indeed, examples from across North America, Europe and Australasia (for example, see Kunkel 2011 on Hamburg; Kerkin 2004 on Melbourne; Ross 2010 on Vancouver; and Hubbard 2004 on Paris and London) suggest that there are many sex-related businesses regarded as incompatible with consumer-fuelled gentrification. In each case, one finds exclusionary metaphors regularly deployed by the forces of law and order as they 'reclaim' inner cities from the 'cultural detritus' of pornography, with commercial sex identified as the lowest of 'all imaginable cultural products', associated with the lowest end of the social spectrum and read as a sure sign of disinvestment (Blomley 2004, 46).

Viewed in conjunction with theories of urban neoliberalism that emphasize the importance of encouraging corporate developers to invest in declined urban areas, this planned displacement of sex businesses can be seen as integral to strategies of re-aestheticization and gentrification (Hubbard 2004). Yet such ideas make less sense in relation to the pornification thesis that argues society is becoming more sexualized and sexually explicit. As such, and acknowledging that many forms of adult entertainment are now imbued with fashionability and profitability, it appears misleading to theorize a wholesale 'zoning out' or displacement of sex businesses given there are many instances where these are becoming more established within the central city. Instead, it appears more appropriate to talk of planning and licensing as attempting to differentiate between sleazy and poorly run business and those perceived to be 'up-market' and potentially attractive to middle class consumers and business travellers (for example lap-dancing chains such as *Spearmint Rhino* and *For your Eyes Only*, which are often represented as the antithesis of the 'bump and grind' strip club). Frank (2008, 65) argues such upscale businesses offer a 'fantasy of distinction', and normalize sex consumption by offering adult entertainment in an environment that can offer fine-dining, charity events, conference rooms and 'a distinguished atmosphere' (see Figure 6.2). The branding of these clubs as 'gentleman's clubs' underlines their aspiration to be recognized as high-end leisure spaces.

But even if (some) local authorities and city governors regard upscale adult entertainment as contributing to the vitality and vibrancy of the city, it is notable that nearly every time a new striptease or lap-dancing club opens, it provokes a wave (or, at the very least, a ripple) of complaints, bringing together local residents, business owners, church groups and anti-pornography groups in sometimes unlikely alliances. While some campaigners put pressure on property owners to refuse leases to clubs, it is more usual that opponents implore local government to prevent clubs opening using the zoning and licensing powers at their disposal. The typical argument here is not that sex businesses should be banned outright, but that clubs can 'pollute', 'taint' or 'contaminate' other land uses such as schools and churches, and are simply not appropriate in 'family' areas. Such 'Not in My Backyard' (NIMBY) protests can lead to protracted arguments in the courts, where the potential negative externalities that sex businesses might have are considered (McCleary and Weinstein 2009). In the UK, where local authorities must consult local

Figure 6.2 *Spearmint Rhino* 'gentleman's club' Birmingham, UK (source: Rhinoedit/GNU Free Documentation Licence).

residents about all licensable premises, including pubs, clubs and cafés, they can refuse a licence on grounds of reducing crime and disorder, promoting public safety, preventing public nuisance and protecting children from harm (see Hadfield and Measham 2009). The idea that lap-dancing venues are inappropriate near to facilities used by children has hence been emphasized in many campaigns of opposition, the latter appearing a particularly effective strategy despite a lack of evidence that the clientele of lap-dancing venues pose any danger to children. Fears about the moral corruption of children and young people have indeed convinced licensing officers that a licence for lap-dancing would not be appropriate in the vicinity of schools or colleges, albeit that in some cases licensing conditions have simply suggested that clubs can be in the proximity of such facilities so long as their opening hours do not coincide with times when young people would be around (Hubbard 2009).

The fact that local residents can object to the presence of lap-dancing clubs in their neighbourhoods only on environmental rather than moral grounds was one factor apparently encouraging a 'boom' in strip clubs in the UK in the early 2000s, with the number tripling from around 100 in 1999 to more than 300 in 2006 (Colosi 2010). The case for tightening control on such venues gained considerable momentum in 2007 when the national anti-pornography campaign group *Object* launched a campaign 'Stripping the Illusion' which argued for increased control of adult entertainment on the basis that it is of a different nature than, for example, live music or karaoke. The subsequent inclusion of clause 27 of the 2009 Crime and Policing Act, which created a new licensing category of Sexual Entertainment Venue, was hence a victory for campaigners and provided a tool for local authorities to refuse a licence for lap-dancing venues on moral grounds, and not just concerns of public order or safety. In effect, adult entertainment venues in the UK are now licensed in the same manner as sex cinemas and 'sex encounter' venues, meaning local authorities will be able to refuse a licence for a lap-dancing venue if it is regarded as inappropriate in the locality, providing a basis for a strict exclusion of striptease venues from specific localities. It also allows stringent conditions to be imposed on clubs, with the possibility of no-nudity and no-touching clauses constraining the entertainment offered, shielding further the already 'hermetic' body of the striptease performer and curtailing what Uebel (2004) terms the 'lure and power of erotic dancing'.

Rather than being viewed merely as a mechanism for reducing perceived environmental nuisance, licensing in the UK has thus been recast as a means to control the form and content of adult entertainment, and is becoming an arena where questions of sexual morality, taste and censorship are blatantly to the fore. Whether or not planning or licensing law should be used to intervene in this way is a moot point, with the new powers having been contested by burlesque dancers as well as others who feel the new legislation will not genuinely differentiate between well-run spaces of entertainment and exploitative or inappropriately located venues, but will simply allow local authorities to deploy stereotyped assumptions about the nature and impact of such clubs without due consideration of their actual effects on local land uses and residential amenity (Hubbard 2009).

Not in front of the children: where should sex shops be located?

Alongside the mainstreaming of lap-dancing venues, the increased prominence and visibility of 'sex shops' is one of the most frequently cited symptoms of the 'sexing up' of society. Although we have already seen that media and representations designed to sexually arouse have long circulated in the city, it was as recently as the 1960s that dedicated stores emerged to sell items that had previously been sold elsewhere (e.g. in pharmacies, specialist bookshops, lingerie stores and via mail order). Though stores were few in number to begin with, and isolated to major cities, they instantly attracted the attention of the press, which in many cases accused them of defying social convention by bringing 'erotic reality' into the realms of the everyday, effectively taking sex out of the bedroom and into the public realm. Tellingly, academic discussions of sex retailing were played out solely within 'deviance studies', explicitly figuring patrons as a specific subset of the population and helping to construct stereotypes of sex shops as only catering to the middle-aged male 'dirty mac' brigade (Tewksbury 1990). Hence, by the time Beat Uhse's sex shops had spread all over West Germany in the late 1970s, and David Gold's *Private Lines* had targeted all UK cities with populations of more than 100,000, such stores were widely opposed by local populations who accused them of encouraging promiscuity and threatening 'family values' despite the frequent marketing of goods as 'marital aids' (Kent and Brown 2006, 200).

Such exclusionary discourses stress that sex shops transgress many of the divides that structure dominant modes of social intelligibility – moral–immoral, high–low, public–private – by taking sex out of the bedroom and into the public realm. Responding, the state sought to introduce forms of legislation to ensure that when sex shops were allowed to exist, they were strictly segregated, enclosed and located so that they did not disturb public sensibility. In some jurisdictions, however, it proved difficult for the state and law to ensure this ordering given operators often argued that the products they sold were permissible within the law and not capable of corrupting adults. Sides (2006) argues that, in US cities at least, pornographers also benefitted from the lesbian and gay liberation movements which had argued that unfettered sexual expression was a fundamental right (see Chapter Two), with the increasing sexual liberalism of the 1960s and 1970s leaving city governors impotent to shut down adult stores on the basis of the obscenity of their stock. Many US cities were hence not able prevent adult retailers and video stores opening in areas zoned for retail uses, and new sex districts began to emerge based around such retailers:

> First in a trickle, then a flood, by the mid-1960s pornography and sex entertainment had moved into the visual life of parts of cities previously untouched by sex districts. Though no longer anchored by the brothel, adult (primarily heterosexual) pleasure zones nevertheless brought commercial sex literally onto the street corner and into the magazine stands, bookstores and local theatres. The presence of commercial sex came to be read through two definitive and powerful middle-class anxieties: threats to the heterosexual, patriarchal family-unit as the organisational foundation of the sexual system, and the changing class character of neighbourhoods.
>
> (Self 2008, 294)

Self's detailed account of Los Angeles – where, by the mid 1970s, there were at least ninety pornographic bookstores – suggests it was not until the legality of zoning ordinances brought in against sex cinemas and theatres in the late 1970s was proven that the city was able to pass zoning ordinances designed to disperse sex shops. Given these ordinances were based on the idea that the presence of sex shops lowered property prices and could promote street crime and drug dealing, civic leaders were then able to take action without being seen to promote censorship.

Similar quandaries have faced civic leaders elsewhere, where censorship and obscenity laws have been used to exercise some control over the content sold in sex shops, but have not been able to prevent these shops opening. Given the legality of the majority of their stock, the most many city governors have been able to do its to limit such premises to non-residential sites, or away from prime retail sites. For example, in most Australian states, stores selling or exhibiting 'materials, compounds, preparations or articles which are used or intended to be used primarily in or in connection with any form of sexual behaviour or activity' are described in planning terms as *restricted premises*, and are deemed to be inappropriate in most residential and office areas because of their lack of capacity to 'engage with the street' given the nature of the goods sold in such premises. When they are permitted, there can be strict constraints on their design and opening: for example, in central Sydney, shops are not allowed within 75 metres of one another lest this creates visible 'clusters' of sex-related land use, and are not permitted to have frontage or window displays at ground level (see Figure 6.3). In France too, a de facto zoning has been applied by virtue of laws that regard sex shops as inappropriate near schools and colleges (see Case Study 6.3).

In the UK, sex-related land uses have never constituted a distinct category in the Use Classes Order, meaning there has been no possibility of using zoning or planning powers to regulate such stores. As sex shops began to appear in particular areas in the 1970s (most notably, Soho in London),

Figure 6.3 The sex shop as 'black box': sex shop exterior, central Sydney (source: author).

concerned residents' groups argued for new legislation, noting that the 1959 Obscene Publications Act no longer justified police raids on sex shops and seizure of material deemed obscene. Given the obvious limitations of both criminal and planning law for controlling sex shops, a number of local authorities followed suit, petitioning central government for powers to exercise control over sex businesses, initially by proposing a change to the Use Classes Order (Manchester 1986). Gradually, however, the notion that licensing might provide a more effective basis for control emerged, given this

CASE STUDY 6.3

A modern moral geography: sex shops in Paris

Paris arguably gained its reputation as a centre for commercial sexuality in the nineteenth century, with 'the bright artificial light of Paris by night stimulating the fantasies that sprang from that milieu' (Corbin 1990, 98). In the poetry of Baudelaire, and later the writing of urbanists like Pierre Mac Orlan and Francis Carco, this reputation was written onto, and out of, the city's *quartiers chauds* (hot districts) including Montmartre, Pigalle, Montparnasse and Rue St Denis. Initially infamous for prostitution, it is the sex shops of these districts that now give them their sense of mystery, danger, adventure and excitement, their red neon signs heralding the fact these are the city's 'erogenous zones' (Redoutey 2009, 191).

According to Coulmont, the first French 'sex shops' opened in 1969 and 1970 at the centre of touristic, civic and intellectual Paris (e.g. the Quartier Latin and the Champs-Élysées). They sold objects that were legal (small vibrators, condoms, various novelties) as well as posters, small artistic statues and books deemed to be unfit for minors but sellable to adults under certain rules. But even if, separately, such objects were legal, their gathering in a single outlet created tensions and concerns that the character of 'old Paris' was being overwhelmed by an avalanche of pornography. For example, in 1969 the conservative daily *Le Figaro* proclaimed 'Eroticism is threatening Paris!' when a store opened under the name '*Sexologie, insolite*'. Elected Parisian officials angrily voiced their opposition, asking the Prefect for 'extremely drastic, forceful measures' against sex shops.

The consequence of such opposition was a two-pronged legal and administrative action. In September 1970, an order (*ordonnance*) of the Paris Prefect created a regulation for bookstores that openly advertised their sexual goods. This '*ordonnance*' created a new class of stores on which the police could act. To counter the weakening of powers banning items deemed offensive to public morality, the Prefect based this decree on another set of laws: namely, the 1949 regulation that created a set of books restricted to adults. The Prefect accordingly adopted the principle that if certain stores specialized in the sale of books forbidden to minors, then such stores should also be off-limits for minors. Further, because sex shop windows became increasingly 'graphic' with the progressive liberalization of magazine content, the Prefect ordered the blacking out of sex shop windows in 1973 so that minors would not be able to see what was for sale. But while the aim was to hide the stores' paraphernalia and restrict it to the private realm, it conversely lent the stores themselves heightened visibility as they were the only ones without real windows. In so doing it created something of a mystique around sex shops, and created them as taboo – and potentially erotic – spaces.

The core of French regulation of sex shops was accordingly protection of the child. This was made explicit in political debates, with one MP arguing that while the law 'forbids the implantation of bars and cafés too close to schools, the implantation of sex shops [is] certainly more dangerous than that of cafés and pubs for the moral security of children' (cited in Coulmont 2007). From the late 1970s, French politicians increasingly argued for tighter regulation of sex shops on this basis, and from 1987, new stores were forbidden within 100 metres of any school. In 2007, this was extended to 200 metres, a ruling that imposes considerable restrictions on the possible locations for sex shops given the sheer density of schools in central Paris.

In the wake of such legislative changes, store owners have been at pains to avoid their shops being classified as sex shops. Making the stores 'couple-friendly' or 'women-friendly' is now commonplace. Coulmont also notes a refusal to forbid entry to minors, facades designed to appeal to upper class tastes, tasteful window displays, the recruitment of professional interior designers, the use of music to induce particular ambiences and so on. While such strategies depend on the legal uncertainty as to what is defined as pornographic and unsuitable for minors in a society where sexual morality is constantly shifting, this is changing the appearance of the *quartiers chauds* as the long-established peep shows and video shops slowly make way for sex supermarkets and self-styled 'love shops'. Given Louis Chevalier, author of the famous work decrying the post-war modernization of Paris (*The Assassination of Paris*, 1984) once accused sex shops of killing love in the 'city of romance', this gentrification and feminization of sex shops in the city is a notable development indeed.

Further reading: Coulmont (2007)

---

had been the main mechanism by which locally contentious land uses such as spaces of gambling and drinking had been regulated in Britain as far back as the nineteenth century. It was during the second reading of the Local Government (Miscellaneous Provisions) Bill – which contained provisions dealing with the licensing of night cafés, tattooing and ear-piercing parlours – that MPs raised the possibility that such licensing be extended to sex shops.

The strength of feeling was such that the government brought forward amendments at the report stage, effectively introducing a system of local sex shop licensing through the provisions of Sections 2 and 3 of the Local Government (Miscellaneous Provisions) Act 1982 (Hubbard *et al.* 2009).

Sex shops have thus been licensed in the UK since 1982 if they sell 'a significant degree of things provided for the purpose of stimulating or encouraging sexual activities' (cited in Kent and Brown 2006, 205). Though the 1982 Act allows local authorities little discretion over the nature of materials sold in sex shops, it permits the imposition of conditions prohibiting the display of products in shop windows (by insisting windows are blacked out) and refuses access to under-eighteens. The 1982 Act also contains provisions that have been used to reduce the visibility of sex shops in areas where 'vulnerable' populations might be, and away from children in particular (Coulmont and Hubbard 2010). Indeed, in the UK it is possible for a local authority to refuse a sex shop licence if it judges that a new licence would exceed the appropriate number in the relevant locality (where an appropriate number might be zero). Grounds for refusing a sex shop seem to be if it is in the locality of facilities used by children and 'families', places of worship and schools. This still leaves the vexed question of how extensive a 'locality' might be, and while it has been established in case law that there may be many localities within a given town or city (so that a locality cannot be defined as being an entire town), localities have been defined variously as approximately 'one-quarter of a mile' and 'one-third of a mile' around a sex shop.

For all of this, the traditional distinction between a shop requiring a licence and one that does not has been the sale of (hard-core) pornographic videos or DVDs classified as R18 (adult only). In an increasing number of cases when a sex shop licence has been refused, a shop has opened anyway with owners simply ensuring that they have no DVDs certified as R18 and that the majority of their stock does not encourage sexual behaviour (claiming that, for example, underwear and leather goods are adult-orientated rather than sexually orientated). As such, and perhaps inadvertently, UK licensing has effectively encouraged a move away from stores focusing on R18 videos and DVDs and led to the increased visibility of (unlicensed) stores selling sex-related goods on the UK high street, such as the *Ann Summers* chain which boasts upwards of 130 shops in the UK selling sex toys and lingerie (see Figure 6.4). Alongside other adult retailers such as *Shhh!* and *Nice and Naughty*, *Ann Summers* is marketed as a women- and couple-friendly store, effectively sold as a 'love shop' as opposed to a sex shop (Smith 2007a). Because these outlets are not legally understood to be sex shops, there is no way for them to be controlled through the licensing process, even if they might appear to

Figure 6.4 ***Ann Summers*** **store, London (source: author).**

have 'a significant degree' of sexual goods for sale (including sex toys, lubricants and sex magazines).

Openly trading on British high streets, this new generation of retailers have effectively gentrified sex retailing, and have arguably made a visit to a sex shop feel more normal. Elaborating, Clarissa Smith (2007a) argues that British sex shops licensed in the 1970s and 1980s remained 'oases of ugliness' whose blacked-out windows created a fear of what lurked within for many female consumers. In her view, this encouraged a furtive and anonymous consumption of pornography, and reproduced these spaces as male preserves. In contrast, the more open and often unlicensed sex shop now challenges the traditional notion of pornography as a male domain. Commenting on unlicensed 'concept' sex stores aimed at women, Kent and Brown (2006, 205) also argue that 'the new female focus on sex-shops has altered both the design of shops and the products they sell'. Described as women's erotic emporia, 'new-style' sex stores sell sex toys and accessories in a 'relaxed and unpretentious environment, where staff are happy to offer advice over a cup of tea', with the interior imagined as female 'playspace'. While some commentators argue these spaces are not sex shops at all, as they are less about sex and more about fashion and indulgence (see, for example, Evans *et al.*

2010), Smith (2007a) concludes that they are important sites in the negotiation of new female sexualities and that their focus on clothing and sex toys is indicative of a feminized hedonism that challenges many existing assumptions about women being 'passive' sexual consumers.

Conclusion

As an articulation of erotic *excess*, pornography has often been viewed by civic leaders as a threat to urban order. Its very existence – as with other forms of abject sexuality – constantly troubles as its exclusion can never be fully achieved; there is an ever-present tendency for pornography to 'leak' out into urban public spaces. For such reasons, it is unsurprising that the authorities have often sought to regulate spaces of pornography and adult entertainment, segregating them from the spaces associated with ordered, monogamous heterosexuality so that its transgressive potential is curtailed while its commercial potential is enhanced. In this chapter, it has been suggested that this process has resulted in the formation of distinctive sexual spaces 'perceived alternately as sites of perverse hedonism, female degredation, benign bacchanalianism, sexual liberation, contagion and violent crime' (Sides 2006, 375).

This chapter has explored these innate tensions and contestations, suggesting that much-maligned sites like sex shops and lap-dancing clubs are becoming more accepted as the acknowledged demand for commercial sex draws consumers and tourists towards cities. Indeed, this chapter has described how many sex-related businesses have been re-dubbed 'adult entertainment', and centralized within the vibrant twenty-four-hour city centres that especially appeal to the creative and gentrifying classes (Bernstein 2007). Brents and Hausbeck (2007, 102) thus identify the proliferation of 'upscale' adult entertainment venues at the heart of Western cities as one of the clearest manifestations of the transformation of the sex industries from a 'small, privately-owned, illegitimate and almost feudal set of businesses dependent on local sheriffs looking the other way' to a 'multi-billion dollar business dominated by corporations'.

For all this, it is clear that places and spaces of sex consumption often remain caught up in the politics of NIMBYism, with those running sex businesses facing opposition from community and resident groups who suggest these are incompatible with the maintenance of social and moral order. In theory, planning and licensing offer an effective basis for mediating such disputes, weighing up varied claims to urban space to decide where sex businesses ought to be allowed. Yet on the basis of this chapter, it might be concluded that both are applied selectively and on the basis of 'common sense' ideas about where sex businesses of different types belong. Lacking clear evidence that sex businesses bequeath negative impacts on their surroundings, or lower property prices, planners and licensing officers often deploy stereotyped views of the owners and clientele of adult venues, and can easily fall into the trap of anticipating that corporations – like *Spearmint Rhino*, *Hustler* and *For Your Eyes Only* – will open the best-managed venues, and those most significant in terms of local job creation. As such, regulation appears to be promoting a particular version of commercialized sex, and far from celebrating the *polymorphous perversity* of the erotic, it could be concluded that the venues most often at the heart of cities are those based on a narrow, but profitable, model of heteronormative fantasy.

Further reading

Bernstein (2007) situates the consumption of sexual services (including prostitution) in the wider context of shifts in intimacy in post-industrial society. Her emphasis on the increased involvement of the middle classes in both the production and consumption of commercial sex offers a useful rejoinder to stereotypes of sex work as associated with abject or residual populations.

Sides (2009) provides a detailed case study of the changing sexual landscapes of an 'erotic city'. His careful excavation of San Francisco's go-go bars, peep shows, sex shops and gay clubs usefully highlights the ways conflicts between pornographers and pressure groups have resulted in distinctive 'sex districts', noting the exclusionary pressures that have pushed commercial sex away from the most valued spaces.

Weitzer (2010) is an edited collection that provides a useful summation of trends in various sectors of the sex industry, including both direct and indirect sex work. The collection is particularly useful on the putative mainstreaming of sex work and the growth of the global sex industry.

7 World cities of sex

Learning objectives

- **To understand how global flows of bodies, commodities, images and ideas are sexed and sexualized.**
- **To appreciate how business, intimacy and sexuality entwine in world cities.**
- **To comprehend the varied ways in which world cities organize the contemporary network of sexualized flows.**

In previous chapters, the focus has primarily been on what happens *within* cities. In contrast, this chapter considers what moves *between* cities: the global flows that bequeath them their increasingly international character and outlook. Recognizing that some cities are more global than others, and seeking to understand the nature of global connectivity, has been a key characteristic of a specific strand of urban research in recent decades. Conventionally, such research identifies world cities as the global hubs that organize ever more complex flows of information, money and people. Yet it is the role of advanced producer services, transnational corporations and financial institutions that tends to be considered as of crucial importance in shaping these flows, often to the neglect of the other cultural and social practices that give world cities their distinctive character. This chapter hence redresses this balance by focusing on sex as one of the drivers of the global economy, arguing that world cities are not merely major markets for sexual consumption, pornography and prostitution but are the hubs of a global network of sexual commerce around which images, bodies and desires circulate voraciously. As such, this chapter answers Short's (2004) call to bring the body into discussions of globalization not merely as a vector of disease transmission, an agent of cultural diffusion or a repository of tacit business knowledges, but as a sexualized body whose intimate geographies are integral to the reproduction of global economic systems which thrive on the commodification of desire.

Given the recent corporatization of the sex industries noted in Chapter Six, this chapter begins by prising open the complex relationships that exist between travel, business, urbanization and sex, noting that many forms of tourism, business travel and recreation involve the purchase of sex and intimacy. Though sex tourism is often depicted as involving flows of Western men to the East, and the movement of sex working women in the opposite direction, this chapter suggests that these stereotypes obscure a more complex global economy of sex. This is an economy that often appears disorderly and amorphous – thanks in no small measure to the rise of virtual porno(geo)-graphies – but which remains grounded in, and regulated via, major world cities. As such, this chapter draws on a multitude of case studies, including Bangkok, Budapest, Dubai, Havana and Las Vegas, to argue that sex underpins the making of world cities at the same time that these world cities articulate an ever expanding global economy of sex and intimacy.

Sites of seduction: in what ways are world cities sexualized?

While the concept of *globalization* is, in comparative terms, relatively recent, the idea that there is a cadre of elite cities with global reach and influence has a more lengthy history. Indeed, the term 'world city' was first coined by Patrick Geddes in 1915 to describe cities in which a 'disproportionate' amount of the world's trade was carried out. Subsequent commentators on world cities have obviously refined this in various ways, whether by highlighting these cities' role in articulating virtual flows, considering them as embedded epistemic business communities or suggesting that their evidential power is related to the thickness and density of institutions they possess (see Hubbard 2006 for a review of world city debates). Slowly, however, and drawing sustenance from emerging ideas about the relational nature of space (e.g. Massey 2006, 2007), world cities are being conceptualized not so much as powerful cities in the world, or simply cities that connect to the world, but as worlds in themselves, whose spatialities challenge traditional notions of local and global:

> In a world dominated and controlled by networks of interaction and global flows, many of the assumptions upon which the territorial conception of cities was founded are increasingly fading away. . . . the key break with the past lies with the fact that while previously the city was thought of as a primary, taken-for-granted entity – stable in time – it can now only be

> envisioned as *one* possible, deliberate construction: a local geographical order born out of the turbulence of global flows and with which it must interact in order to continue to exist.
>
> (Dematteis 2000, 114)

Notwithstanding the dominance of this topological and relational way of thinking of world cities, there has remained a tendency to identify a discrete set of metropolitan centres (London, New York and Tokyo) that are truly global cities in the sense that they exist, and are produced, everywhere. Beyond these, it is postulated that there is a second tier of cities which articulate specific 'globalization arenas' (e.g. Los Angeles in North America, Paris and Frankfurt in Europe), and beneath that a third tier of cities of somewhat lesser importance, such as Seoul, Madrid, Sydney or Singapore. Significantly, the basis for constructing such hierarchies of world cities has typically been the economic attributes of particular cities for which data are readily available to researchers in the urban West (e.g. the number of corporate headquarters, stock exchange activity and inter-firm flows of communication). While such indicators apparently reveal the most globally dispersed and influential world cities – rather than those which merely boast a large population – we need to be mindful of the limits of such statistics, which prioritize particular forms of work over others, and generally ignore the embodied dimensions of urban life. The world city network is, after all, made not solely through flows of money, ideas and information, but through flows of people whose decisions to move or migrate are shaped by their sexualities in a number of important ways. Yet in those accounts that have sought to reintegrate migration in the study of world cities (Malecki and Ewers 2007) there is rarely any consideration of sex, love or intimacy. Noting this omission, Mai and King (2009, 297) insist that 'beyond their common function as mobile workers within the global capitalist economy . . . migrants and other "people on the move" are sexual beings expressing, wanting to express, or denied the means to express, their sexual identities'.

But how might studies of global mobilities incorporate questions of love and sex into migration analysis? One study that begins to do this is Walsh's (2007) examination of the heterosexualities performed by young British expatriate workers in Dubai. Having massively expanded since the discovery of oil in the 1960s, Dubai has become one of the most important cities on the Arab peninsula, its emergence as a world city associated with its growing role as a centre for financial services as well as its reputation as a luxury tourist resort (see also Case Study 4.1, Chapter Four). Significantly, it is also a city where 80 per cent of the resident population are transnational migrants, including both low wage workers employed in construction and tourism work and higher paid workers in finance and advanced producer services. Focusing on the latter, Walsh reports that there is a tendency for young British expats in Dubai to reject notions of coupledom and instead perform a transient heterosexuality focused on play, freedom and the pursuit of sexual pleasure. Significantly in a city where the consumption of alcohol is not publicly tolerated by the indigenous Emirati, this means that a series of hybrid bars/clubs have become important in the social lives of the expats, with dressing up, dancing and flirting being almost daily activities for many of them. Walsh found that most of the young (heterosexual identified) expats she interviewed saw their time in Dubai as something of a working holiday, behaving differently than they might when in their 'home' country: in the case of Dubai this *liminality* was emphasized by the climate and beaches that made Dubai feel like a holiday space to the expats.

To suggest that some transnational workers perform different sexualities when working in foreign contexts is to highlight the importance of sexuality both in decisions to migrate and in migrant lifestyles within world cities. It is also to underline that sites of sociality and sexuality ('landscapes of desire', as Walsh terms them) are integral to world city formation, providing spaces where transnational elites can cement friendships with business contacts and work colleagues. While some of these have been identified in literatures on gay tourism (e.g. see Ingram's (2007) discussion of the gay life of Dubai's beaches), the sexual dimensions of expatriate and migrant lives are largely ignored in studies where heterosexual identity positions are assumed, notwithstanding that much work on transnational business masculinities notes the importance of the bonding played out in sexualized spaces (e.g. Beasley 2008). It is no coincidence, therefore, that leading world cities are known for their red light districts and sexual entertainment aimed at business visitors as much as at 'local' consumers. For instance, districts at the heart of London (Soho), Amsterdam (Wallen), Paris (Pigalle), Tokyo (Kubukicho) and New York's Times Square district enjoy internationally mediated notoriety for sexual commerce, their reputation enhanced by the proliferation of guidebooks, brochures and (especially) websites which provide guides to the red light landscapes of these cities. Herein, the sexual possibilities of the metropolis are mapped out in sometimes bewildering detail: the *Paris Sexy* (2004) guide, for instance, offers fourteen chapters of advice as to where travellers may locate sex workers, BDSM dungeons, swinger's clubs, saunas, striptease bars, pornographic cinemas, cabaret and other '*spectacles d'érotisme*'. Sometimes, this form of sexual advertising is officially sanctioned, with sex work identified as a potential marketing tool (or 'soft location factor') in the global battle for jobs and investment. For instance, the Netherlands Board of Tourism and Convention identifies the Red Light District as a key 'quarter' of Amsterdam, urging a visit:

> From brothels to sex shops to museums, the Red Light District leaves nothing to the imagination. It is very likely that you will have heard about this neighbourhood and to be frank, everything you will have heard is probably true, but to really put rumours to rest, you have got to check it out for yourself. The *Rossebuurt*, as the locals know it, is unlike any other place. Guaranteed. Certainly, the Red Light District that everyone knows about is the one where women, of all nationalities, parade their wares in red-fringed window parlours, many ready to offer more than a schoolboy peep-show in a private cabin.
>
> (www.amsterdam.info/red-light-district/ accessed May 2011)

Significantly, the same source recommends visitors locate the area's 'infamous' condom shop before commencing their trip, and notes that taking photographs of working women is bad etiquette (see also Aalbers 2005). Other cities also make much of their 'sexy' reputations in tourist promotion, with the emergence of sex museums as visitor attractions a notable trend (see Figure 7.1).

Whether or not governors of world cities encourage such representations in the interest of cultivating a 'sexy' reputation for their city is open to question. However, the connection between sex and economic growth in the post-industrial city has certainly been made explicit in the work of globalization 'guru' Richard Florida. Put simply, his thesis (as proposed in *The Creative Class*) is that cities require a critical mass of creatives to thrive, a class of workers whose job is to create meaningful new forms of work (Florida 2002). This 'class', which Florida estimates to consist of some 150 million individuals worldwide, is composed of scientists and engineers, university professors, people in design, education, arts, music and entertainment, whose economic function is to create new ideas, new technology and/or creative content. According to Florida, for a city to become a magnet for the Creative Class, it must be an example of 'the three Ts', providing Talent (have a highly talented, educated and skilled population), Technology (have the technological infrastructure necessary to fuel an entrepreneurial culture) and Tolerance (having a diverse community with a 'live and let live' ethos). One way that Florida operationalizes the latter is by a diversity index based on the proportion of coupled gay households in a region, seeing this as a good predictor of creativity and urban productivity:

> Gays, as we like to say, can be thought of as canaries of the creative economy, and serve as a strong signal of a diverse, progressive environment. Indeed, gays are frequently cited as harbingers of redevelopment

Figure 7.1 New York's Museum of Sex (source: David Shankbone/GNU Free Documentation License).

> and gentrification in distressed urban neighborhoods. The presence of gays in a metropolitan area also provides a barometer for a broad spectrum of amenities attractive to adults, especially those without children.
>
> (Florida and Gates 2005, 135)

Deploying such logics Florida identifies a clear relationship between this gay index and the relative output of an area's high-tech industries, noting that eleven of the top fifteen high-tech metropolitan areas in the US also appear in the top fifteen of his gay index, with San Francisco topping both lists. Even when removing San Francisco from his analysis (because it ranks 'unusually high' on both measures), he concludes that a metropolitan area's percentage of gay households remains the only significant predictor of high-tech growth even when other regional measures of diversity, such as percentage of population born overseas, are factored in (Florida 2002).

The idea that gay populations are particularly creative and/or artistic is a widespread myth, but one that numerous gay and lesbian organizations have been keen to play up (see Forest 1995). Nonetheless, Florida's ideas have been subsequently critiqued on both empirical and conceptual grounds (e.g. Marlet and van Woerkens 2005 find that the 'gayness' of Dutch cities, measured in subscriptions to gay magazines, is a poor predictor of the

CASE STUDY 7.1

Promoting world cityness through sexy tourism: marketing Gay Cape Town

Though apartheid governments effectively used the regulation of sexuality as a means of disciplining those who opposed apartheid (Conway 2004), in 1996 South Africa became the first country in the world to prohibit sexual discrimination, integrating clauses on the right to privacy in the Constitution. In 2005 same-sex civil unions were legalized. Such progressive reforms represented an effective loosening of the state surveillance of sexuality that was exercised in the apartheid era, and allowed for the formation of more visible gay spaces in South African cities. Tucker (2009) argues that this established South Africa as the most progressive country on the continent, suggesting no other country changed its position towards queer individuals or the world's perception of itself in such a short time. Nonetheless, the distinctive racial and sexed history of South Africa, as well as a context where HIV is widespread (South Africa is thought to have the highest rates of HIV infection in the world, albeit with 80 per cent of cases among women), means that its queer visibilities take particular forms.

The post-apartheid rush to attract foreign investment to the nation has been closely tied to tourism, using sporting spectacles like the Rugby (1995), Cricket (2003) and Football (2010) World Cups as a platform for promoting a distinctive tourist 'offer'. There has been an explicit focus on younger, single men in much tourist promotion given this group are regarded as 'playboys' with sizeable disposable incomes and no family ties. As part of this process, gay male tourists have been targeted, with Cape Town actively sold as the start of a 'Pink Route' across South Africa which incorporates a series of gay-friendly hotels and guesthouses but which takes in national parks, the annual Pink Loerie Mardi Gras in Kyhsha, Annual Gay Pride events, queer film festivals and so on. At the same time, young South Africans have taken opportunities to travel more widely, taking advantage of improving relationships with countries in Europe and North America in particular, bringing back specific ideas of what gay leisure space should consist of. Visser (2003) suggests these factors combined to produce notable foci for gay leisure in South African cities, often centred on the same type of diverse leisure facilities found in 'international' gay capitals like Amsterdam, London and San Francisco. In Cape Town, it was the predominantly white, middle class dance clubs, saunas and restaurants of De Waterkant that have been promoted as the centre of 'Gay Cape Town', overshadowing the township 'shebeens' (unlicensed pubs), bars, parks and beaches that provided the traditional 'backbone' of gay male life in the city (Tucker 2009).

Cape Town is thus self-consciously marketed as one of the world's leading gay cities, with the gentrified De Waterkant acting as a focus. Elder (2005) contends that these processes of gay commodification and marketing have produced a homogeneous 'globalized' form of gay space in this area that bears much similarity to the types of gay village found in Amsterdam, Sydney or London. He further argues that this has relied upon a selective marketing of the city that effaces its apartheid past and plays up merely a colonial 'style', attempting to draw in tourists for whom questions of the oppressive pasts of South African do not exist (or are easily forgotten). Moreover, he shows that the marketing of 'gay safe space' in Cape Town is nothing of the sort, being intentionally male, white and exclusive of individuals who might feel anxious on streets where a homosocial ambience dominates. Visser (2003) similarly concludes that, in a city where only 25 per cent of the population is white, the marketing of Gay Cape Town equates whiteness and gayness in a dangerous manner, marginalizing the coloured and black communities. In such ways, attempts to tie Cape Town into flows of gay tourism have led to criticisms that it is perpetuating a particular 'glocalized' gay identity that is 'cuddly' and non-threatening (Binnie 2004) but which is actually profoundly classed, coloured and gendered. The fact that homophobic violence has been perpetuated within the spaces marketed as safe leads Elder (2005) to conclude the identification of de-differentiated gay spaces is nothing short of a 'scam'.

Further reading: Visser (2003)

---

distribution of creatives in Dutch cities, while Bell and Binnie 2004, 1821, allege that Florida's formulation is nothing more than an index of respectability and 'nicely gentrified neighbourhoods'). The measure of 'gayness' adapted by Florida is certainly highly problematic given that measures of gay and lesbian cohabitation only capture a fraction of those identifying as gay and lesbian; arguably this measure also stereotypes dual-income no-kids households as pursuing consumer-based, aestheticized lifestyles. Despite such critiques, the idea that there may be some connection between sexuality and world city formation is not easily ignored, and certainly feeds into numerous policy initiatives designed to market 'wannabe' world cities as hip and happening gay capitals (Markwell 2002; Puar 2002; Hughes 2003).

This process of marketing cities as 'gay' centres has thus fuelled debates about the colonization of indigenous sexual scenes by a Western-inflected gay or lesbian culture, with Altman (2004, 64), arguing that 'one sees unmistakable signs of American lesbian/gay imagery and self-presentation in almost every part of the rich world' (see also Shahani 2008). It has also highlighted the processes that lead to the construction of a 'global gay' who is assumed to be able to travel easily and smoothly, but is not necessarily open to the cultural experiences that would produce forms of sexual hybridity or difference (Oswin 2006). Nevertheless, global tourism processes appear to be creating a world of competing gay tourist destinations, all chasing the mythical 'pink dollar'. What is perhaps less noted is that an increasing number of city governors have recognized the importance of sexual and adult entertainment in attracting business tourists and conference travellers from beyond the lesbian and gay community (though see Sanchez 2004, and *Case Study 6.1* on Portland, Oregon). Unsurprisingly, few surveys of business travellers ever suggest sex is a motivation for travel, but anecdotal evidence for this is legion, leading numerous commentators to hypothesize an explicit connection between business travel and sexual consumption, especially when such travel allows individuals to escape the confines of an existing, coupled relationship (Wonders and Michalowski 2001). While this can apply to women travellers, much commentary fixates on the (assumed heterosexual) business*man*, who is thought to be a significant and sometimes voracious consumer of sexual services when away from his home country. For example, Marttila (2008) suggests that around 50 per cent of clients of prostitutes in Estonia are Finnish men drawn to the city not so much because of the presence of sex work but because of its pivotal role as a trading 'crossroads' between East and Western Europe:

> A majority of the Finnish men buying sex in Tallinn and Vyborg do not go there exclusively for that purpose, the trip in question often being a business or weekend trip during which paid sex services are used. Most of the Finnish men I encountered in brothels and sex bars in the region did not identify themselves as sex tourists. Many of them were on business trips . . . in downtown Tallinn people often offer to sell Viagra to groups of men looking 'businessmen-like'.
>
> (Marttila 2008, 49)

In Tokyo and Shanghai, visits to massage parlours and hostess bars appear to be as much a part of corporate hospitality as banqueting and karaoke (Allison 1994; Zheng 2009), while in London there have been numerous cases where female employees have complained about the culture of entertaining foreign visitors by taking them to lap-dancing and strip shows (Rutherford 1999). Drawing connections between sex entertainment and corporate cultures, Holgersson and Svanstrom (2004) likewise allege that visits to strip clubs in Stockholm are homosocial occasions where men together confirm their gender identity and superiority over women, developing forms of intimacy between themselves in a space of heteronormative consumption (see also Liepe Levinson 2002).

Beyond this, it is clear that hotels catering to business travellers are important sites for escort work and prostitution, while sexual consumption is normalized in the provision of hard-core porn on the pay-to-view channels that constitute part of the in-room entertainment. As Pritchard and Morgan (2006, 769) argue, the 'very liminality of hotels – as crossing points into the unknown, as places of transition and anonymity, hidden from familiar scrutiny – makes them attractive as venues for sexual adventure'. Unquestionably, this relies on staff imbuing hotels with sexual values through particular forms of bodywork, and sometimes through deliberate strategies of looks-based staff profiling (McNeill 2008). Distinctions between 'love hotels' (see Case Study 3.1) and other hotels/motels are thus becoming hard to discern in an experience economy where sex, adventure and romance are entwined in commercial hospitality (Alexander *et al.* 2010).

All of this is to insist on the importance of sex in the making of transnational business networks, and to problematize any neat distinction between business and sex tourism. Noting this blurring, Oppermann (1999) seeks to reconceptualize sex tourism, arguing the 'ideal' sex tourist – who purposely takes a holiday to have sex, stays away from home for at least twenty-four hours, has sexual intercourse as a result of direct monetary exchange and obtains sexual gratification in encounters which last a relatively short time – usually does not exist. Oppermann continues by asserting that the vast majority of tourists who visit prostitutes or sites of adult entertainment in the pursuit of pleasure do not travel for that purpose alone. O'Connell Davidson (2001) describes these as *situational* sex tourists. Dispensing with clichés in which the sex tourist appears as a predatory and potentially paedophiliac male, travelling solely to exploit and dominate economically subordinate young people, this notion of situational sex tourism forces acknowledgement that sex and business entwine in a multitude of ways (including the purchase of sexual services by women – see Takeyama 2005; Jacobs 2009).

Erotic cities: how is sex implicated in 'globalization from below'?

Acknowledging that existing networks of business and finance are sexualized in various ways provides one perspective on the importance of sex in the making of world cities. But to suggest the sex industries simply follow existing flows of finance and business is to downplay the importance of the sex industry as one of the drivers of the global economy, and fails to recognize that it has its own distinctive geographies. Historically, the sex industry has been associated with financial centres, where sex workers often cater to the wealthy (Sassen 2002), but it has also developed in tourist areas (Pope 2005), mining towns (Laite 2009), transport hubs (Kuhanen 2010), in border zones (Hofmann 2010) and in centres of conflict (such as occupied zones – see Moon 1997): in short, any areas where men have been present but where their normal partners have been absent (Ryder 2004). There are many cities whose importance as centres of sexual commerce is therefore disproportionate to their significance as centres of finance. This implies that while all world cities are seductive, some cities are decidedly more seductive than others, becoming de facto *erotic cities* – a term Sides (2009) coins to describe San Francisco (home of the world's first striptease bar and most famous 'gay village') but which can certainly be applied to other cities.

Beach cities provide one example. Such tourist destinations are often described as eroticized, with the tendency for visitors to wear fewer clothes than they might usually do one contributor to the creation of a sexualized atmosphere (see Simon 2002 on Atlantic City; Hemingway 2006 on the carnivalesque of Brighton; and Lewis and Pile 1996 on Rio). The provision of bawdy, raunchy or adult entertainment can further make sex an important component of a city's tourist economy. Bangkok is a particular case in point. Positioned within East–West flows of tourism, a context of rural deprivation, gendered inequality and a predominantly Buddhist attitude which is tolerant of many forms of sex working, Thailand has been a notorious centre for sex-related tourism since at least the 1960s, when US military en route to Vietnam stopped there for rest and recreation (Askew 1998). In some senses, this reputation is undeserved, for much of the sex work in Thailand caters for local populations, and the country has sought to discourage the package tour 'sex holidays' that were evidentially popular with German and Japanese tourists in the 1980s. But if sex is not the motivation for all tourists to visit the country, Bangkok is still described by many as 'the brothel of the world' where Thai sex workers perform for male clients in a bewildering range of go-go bars, sex shows, karaoke clubs and coffee shops where they do not merely sell sex but a fantasy of an Oriental woman who is not only physically beautiful and sexually exciting but also caring, compliant, submissive and not Western or modern (see Figure 7.2). Between them, the key 'red light areas' of Bangkok – Patpong Road, Nana Plaza and Soi Cowboy – offer over three hundred bars and clubs where workers perform a variety of acts, with some estimates suggesting that there are as many as 200,000 employed in the city's adult entertainment sector (Bishop and Robinson 1998; Babb 2007).

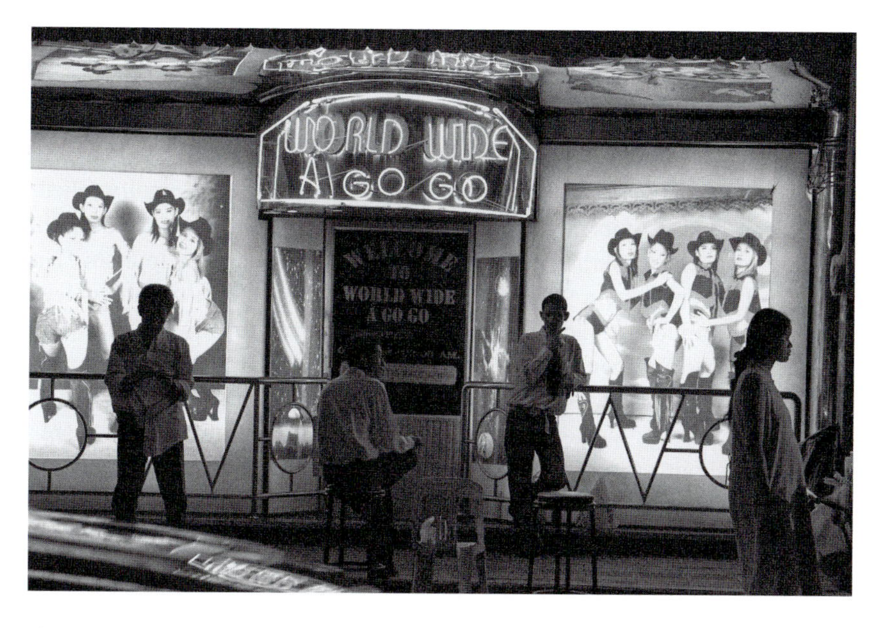

Figure 7.2 ***World Wide A Go Go Club,*** **Bangkok (photo: Kay Charnush for the US State Dept).**

Often condemned as a city where predatory heterosexuality is normalized, and women subordinated, some academic accounts nonetheless challenge ideas that workers in Bangkok are victims, and highlight their capacity to gain significant material and emotional benefits from their interactions with foreign men. Joan Phillip and Graham Dann go so far as to argue that sex workers in tourist bars in Bangkok should be recognized as 'entrepreneurs', involved in the 'ordinary' risk taking activities associated with any business enterprise (Phillip and Dann 1998, 70). Moreover, as a contact zone where different cultures and moralities collide, and where identities can 'oscillate' (Askew 1998), Bangkok represents a city where new eroticized and gendered identities have emerged as a result of the multiple, complex and locally inflected encounters played out there. A case in point here is the infamous Bangkok 'ladyboy', an identity that has its origins in the category of *kathoey*, 'effeminate men' who were tolerated in Thai society but not accepted as de facto women (Jackson 2009). The subsequent ubiquity of ladyboy shows in Bangkok, and the global mediation of these performances, means that *kathoey* identities have become increasingly valued, demonstrating that within a context of global change, new 'placed' identities may emerge that vary just as radically from their own pasts as they do from contemporary Western ideas of idealized gay male identities. The contemporary *kathoey* identity may have taken shape in Bangkok, but is a product of global entanglements of desire (Jackson 1997).

If Bangkok is the sexual playground of South East Asia, Las Vegas can certainly claim that title in North America. A nineteenth-century frontier town, Las Vegas' remarkable expansion in the twentieth century relied upon the popularity of its night-time economy, meaning that it now boasts a unique high wage, low skilled economy based on gambling and tourism. The fact that Las Vegas is in the only state of the US where prostitution is legal has been an important factor in maintaining this growth, allowing Las Vegas to market itself as 'Sin City', a place where sexual fantasies and adult pleasures can come true (Brents *et al*. 2009). While prostitution per se is not (officially) part of the Las Vegas experience, given all the licensed brothels are more than an hour's drive from the city, the resort industry has always relied upon the marketing of sex, embodied in the form of the Vegas showgirl, an idealized and glamourous dancer whose provocative form has entertained visitors since burlesque was introduced to the Strip in the 1950s at celebrated venues such as *Minsky's*, *Stardust* and the *Desert Inn*.

Commenting on the iconic status of the Vegas showgirl, Riechl (2002) contends that the economy of Las Vegas depends not so much on the selling of sex but the *production of desire*, which draws people to the city in the pursuit of pleasure. Gambling, of course, remains the cornerstone of the economy, but even conservative estimates suggest there are more employed in the sex industry than are employed in the casinos. While street prostitution has long since been displaced from the Strip, a few blocks away from the main resort hotels, striptease and gentleman's clubs like *Spearmint Rhino*, *Déjà Vu*, *Sapphire* and *Crazy Horse Too* prosper (with something like 15,000 dancers having the sheriff's licence needed to work such venues). Moreover, as revenues from gambling dropped at the turn of the century, the resort hotels themselves have become more blatant in their marketing as adult venues, with newcomers like the *Hard Rock* casino raising the stakes by marketing itself as a venue where the 'bucking never stops' (Engstrom 2007; McGinley 2007). Increasingly frustrated with attempts by the Convention and Visitors Bureau to downplay the importance of adult entertainment in the tourist experience, in 2002 a consortium of adult entertainment companies formed the Sin City Chamber of Commerce: as well as the legal brothels from out of town, this includes representatives of around one hundred firms active in adult entertainment, including limousine stripping agencies, bachelor and bucks party organizers, sex shops, gentleman's clubs and swinging venues (the latter being an acknowledged part of the Las Vegas sex scene) (Sheehan 2004). The fact that the annual US Adult Video Network Expo is in the city further underlines the importance of adult entertainment for the city (it attracts upwards of 25,000 delegates over three days).

Given that Las Vegas offers a sex-soaked world of lap-dancing, stripping, swinging and big casino topless showgirl revues, it is perhaps unsurprising that up to 3,500 illegal prostitutes work in Las Vegas' underground economy at any given time (Brents and Hausbeck 2007). While some independent workers ply their trade discretely on the casino floor, escort work is much more important, and the majority of hotels appear complicit in arranging 'personal services' should guests request these. Despite efforts by the larger resort hotels to prevent hawkers handing out contact magazines and escort brochures on the Strip (Riechl 2002), stands containing flyers for escorts are found throughout the city (see Figure 7.3), leaving visitors in little doubt that Las Vegas is able to cater for a variety of sexual tastes.

Examples such as Bangkok and Las Vegas begin to imply that if we were to draw up a roster of erotic cities, it would appear somewhat different to conventional mappings and rankings of world cities. Cities like Las Vegas, as

Figure 7.3 Adult 'contact' magazine dispensers on the sidewalk, Las Vegas (source: John Harrison).

well the European 'stag capitals' of Tallinn, Hamburg, Riga, Prague and Amsterdam, would figure strongly. But it is often in emerging tourist areas in the global South that one finds the economy significantly given over to the mixing of relatively wealthy tourist-clients and relatively impoverished sex workers:

> The proliferation of sex-tourist destinations throughout the developing world reflects global capital's destabilizing effects on less industrialized countries' economies. This economic globalization not only shapes women's work options in the developing world but also forces them into insecure, and possibly dangerous, work. In sex-tourist destinations we find the tremendous effects of global capital: its redirection of development and local employment into the tourist industry, especially women's work and migration choices; its creation of powerful images, fantasies, and desires that are linked to race and gender; and its generation of new transnational practices from which foreigners extract more benefit than locals.
>
> (Brennan 2010, 205)

Even in nations that have only relatively recently become open to global flows of tourism and migration, such as China and Cuba (see Case Study 7.2), tourist orientated prostitution has become highly significant, binding specific cities into world city networks of sex tourism. In many cases, these networks perpetuate established global inequalities, carrying traces of earlier colonial eras in which many European powers seemed to view sexual conquest as inherent to projects of modernization and imperial expansion.

The importance of sex work in emerging world cities (such as Havana) suggests that any assessment of a city's importance in global networks needs to take some account of the sex industries as a driver of economic growth. However, the likelihood of being able to acquire reliable data on the size of these cities' sexual economies is slim, as much of what occurs in the sex industries remains in the shadow economy, despite attempts in some jurisdictions to bring it into the mainstream through licensing and taxation (Sanders 2009b). The majority of sex sold worldwide occurs within a context of quasi-legality, so that while the selling and buying of sex is not illegal, it is hard to run a sex business legitimately without falling foul of laws controlling brothel keeping, sex advertising or pimping. Such de facto criminalization creates certain opportunities for exploitation, with workers – including vulnerable young men and women – seeking 'protection' from managers and agents. Penttinen (2004) has accordingly used the concept of *shadow globalization* when referring to the global flows of migrants who are employed in the sex industry. The extent to which migrant workers dominate sex markets varies massively, but there is certainly plentiful evidence to suggest that non-native and/or illegal migrant workers make up the majority of sex workers in major world cities. For example, Tokyo's sex market is depicted as dominated by indentured Thai and Philippine workers, while Amsterdam's licensed windows are mainly worked by migrant workers from Dutch colonies (e.g. Surinam and Antilles). In the UK, some reports suggest 80 per cent of workers in London's off-street brothels and massage parlours are migrants, from Latin America and South East Asia, but principally Eastern Europe.

Given the feminization of poverty, gender discrimination and emerging 'migration cultures' in the countries of origin (Mai 2009), the increasing involvement of migrant women in sex markets is not hard to understand. These migratory factors are not only confined to women, with male sex work in London, for example, significantly populated by migrant men who arrive from places such as South America and Eastern Europe (Gaffney and Beverley 2001). In a contemporary context, there is much concern that a large proportion – and maybe even a majority – of these men and women will have been trafficked for the express purpose of sexual exploitation. This is tied into the identification of trafficking as a significant by-product of the thickening of global networks, with 'the growth of shadow economies and transnational criminal networks' a 'negative manifestation of globalization, arising from expanding economic, political and social transnational linkages that are increasingly beyond local and state control' (Goodey 2003, 417).

Sex trafficking is obviously not new, yet contemporary commentators accordingly suggest it far exceeds the levels that prompted (for example) the formation of a League of Nations Committee in 1933 to address 'a certain movement of occidental prostitutes to the Orient' (Self 2003, 78). Indeed, it is estimated somewhere between 400,000 and 1,000,000 people are trafficked globally annually, with a significant – but ultimately immeasurable – number ending up working as prostitutes (see Hubbard *et al.* 2008). Such estimates are highly suggestive of global patterns of trafficking, but a major issue clouding the trafficking debate is the uncertainty about the proportion of prostitutes who have been trafficked and those who have migrated voluntarily. Particularly problematic here is the distinction between migration via smuggling networks and enforced migration at the hands of traffickers – a distinction many human rights organizations and feminist activists claim is irrelevant given those seeking to migrate with the aid of people smugglers do so in desperate circumstances and with little knowledge about potential opportunities for employment (Hughes 2002). However, others reject this to posit a more complex range of scenarios situated at different points on a continuum of voluntary and involuntary migration. For example, Agustin (2007)

CASE STUDY 7.2

Tourist-orientated prostitution: sex work in Havana

As one of the last bastions of communism, Cuba has considerable draw for Western tourists who wish to experience the 'Other', its vibrant culture, Caribbean climate and recent political history combining to provide a distinctive tourist 'product'. Given tourism has become the most important single source of hard currency, sex work has been a major means by which many Cubans have sought to acquire tourist dollars – an important commodity in a high inflationary economy where the acquisition of dollars has become crucial to secure the purchase of many non-essential goods. In Castro's Cuba, prostitution was made illegal and all brothels closed down in 1959: however, by the late 1990s, Cuba was becoming known again as a centre for prostitution, with this tacitly accepted by the authorities keen to exploit Cuba's reputation as exotic and erotic. Castro himself even defended prostitution in a 1992 speech to the National Assembly in which he noted that sex-working women in the country provided a vital service for tourists, and, moreover, were generally better educated and healthier than those in other nations (Pope 2005).

Havana, as the capital, remains the focus of around three-quarters of all tourist stays in Cuba. Until the mid 1990s, tourists in Havana were limited to designated hotels, and the interaction of locals and tourists discouraged via policy. Subsequent changes from the 1990s onwards, including the legalization of the US dollar as a currency, and the private rent of rooms by tourists, provided new opportunities for sex work, with both male and female workers touting for business primarily in bars and restaurants. This given, sex work shifted from being publicly solicited in the traditional (pre-revolution) red light districts to being offered through third parties, at dollars-only nightclubs, and at certain *zonas de tolerancia* such as in the plazas and streets of Old Havana and the hotel areas of the Vedado neighbourhood.

Described as *jineteras* (female hustlers) or *pingas* (male escorts) rather than prostitutes, Havana's sex workers mainly work independently and entrepreneurially, selling themselves not for sex per se but as pseudo-girlfriends or boyfriends who will offer companionship over the course of a tourist's stay. From the other perspective, tourists who go to Cuba often represent themselves not as sex tourists but 'romance seekers' who lavish gifts and attention on the sex workers rather than simply paying for sex. This suggests that the relationships between hosts and guests need to be regarded as somewhat fluid, and that it is hard to distinguish tourist-orientated prostitution from the search for companionship, with many of the younger workers seeing their liaisons with tourists as a route to possible marriage or migration (Cabezas 2004).

Pope (2005, 112) argues that such emerging geographies of sex work in Havana are a result of 'recent national policies, social practices, revolutionary culture, the spatial organization of society, and increasing interconnectedness among countries'. Yet she also notes that these emerge from 'colonial paradoxes and fantasies/imaginings of the hypersexual yet romantic, liberated yet loving, educated yet submissive, mulatta woman'. Similarly, Wonders and Michalowski (2001, 551) argue that the emotionally attentive and often mixed-race *jineteras* of Havana constitute a 'fetishised' ideal of 'the imaginary hot Latin and the equally imaginary sexually insatiable African'. As such, they conclude that Havana is intimately connected to cross-border circuits of sexual fantasy, and that the maintenance of steady flows of Western tourists to the city is encouraged by the image of the city as exotic and erotic. In this context, Pope (2005) concludes that when the Cuban female *jinetera* body is (literally) penetrated by capitalists, it is a painful reminder both of pre-revolutionary days and of Cuba's current dependence on foreign aid.

Further reading: Pope (2005)

---

argues that many of the immediate opportunities facing migrants in domestic and caring work are carried out in informal and nearly feudal conditions, and that many migrants prefer to sell sex instead, despite the stigma attached to it. In Mai's (2009) study of migrant sex work in London, most interviewees underlined that they enjoyed respectful and friendly relations with colleagues and clients and that by working in the sex industry they had better working and living conditions than those they encountered in other sectors of employment (mainly in the hospitality and care sectors). Tellingly, only a handful (around 6 per cent of female interviewees) in his sample felt that they had been deceived and forced into selling sex in circumstances over which they no control.

Recent ethnographic work on cross-border marriages, domestic work and sex work thus problematizes binary notions of migration being for either love or money, and posits more complex relationships between transnational intimacies and the commodification of personal relations (Constable 2009). In spite of this, the conflation of migrant sex work and trafficking continues, and continues to inform governmental policies that view the presence of migrant workers in the sex industry as evidence of 'traffic in women'. Trafficking panics, such as those concerning the potential rise in sex trafficking associated with major urban spectacles and sporting events such as the 2006 World Cup in Germany, the 2010 Winter Olympics in Vancouver and 2010 World Cup in South Africa, have proved to be largely unfounded (see Hennig 2006; Bird 2009) but this has certainly not prevented the introduction of highly symbolic acts designed to prevent the 'penetration' of Western cities by a seemingly unstoppable influx of 'Eastern girls' (Berman 2003).

Porno-economies and world cities: what role do world cities play in the production and distribution of pornography?

So far we have seen that globalization is implicated in the making of new migration flows between cities. Yet for many commentators on global affairs, one of the most important aspects of globalization is the amount of social interaction – and business – conducted at a distance. The Internet is a key technology here, and has been particularly important in the context of the sex industry, with pornography producers being early innovators in the use of online technologies to disseminate adult content. This has enabled online users to access a range of materials and images that might otherwise have remained hard to access because of legal, moral and social sanctions against them. Even in nations where governments have sought to block adult content and prevent peer-to-peer file sharing of pornography, it remains relatively easy to access sites where sexual imagery can be freely viewed (and posted), while there are thousands – perhaps hundreds of thousands – of sites where video content is available to download on a subscription or a pay-to-view basis. Such sites have fundamentally transformed the nature of pornographic consumption, effectively *domesticating* pornography. Long-standing attempts to restrict the visibility of pornographic materials by enclosing them within closely surveyed (male) realms, such as the sex shop or licensed sex cinema (Coulmont 2007; Mikkola 2008; Hubbard *et al.* 2009), have hence began to unravel as sexual content has become increasingly integrated within the public sphere of the Internet, prompting moral panics about the erosive effects of pornography on domestic and family life (see Chapter Four).

For such reasons, the world wide web is often described as complicit in the pornification of society (see Chapter Six): industry statistics suggest that around one-third of Internet users are now regular visitors to online porn sites, with free-to-view porn websites (such as Pornhub, the sixtieth most popular website in the world in February 2011, and YouPorn, the sixty-seventh) among the most visited sites on the web. In addition, subscription-based porn and adult dating sites create an annual revenue in excess of four billion US dollars (Edelman 2009). Such figures may well be unreliable given the tendency for the adult industry to exaggerate its own importance, but are certainly indicative of the importance of pornography as a virtual business given it accounts for around 70 per cent of all purchases of online Internet content, outstripping sports, news and video games. What is perhaps more significant in financial terms is that many of the estimated fifteen million web pages with adult content offer opportunities to purchase 'real' pornographic content (in the form of DVDs and magazines, sex toys or erotica), and encourage subscription to online webcam and sex chat services, placing the Internet at the apex of a global pornography industry worth as much as ninety-seven billion US dollars worldwide in 2010, with only music and games more important as sources of revenue (D'Orlando 2011). Online sex is clearly big business.

For some, Internet pornography is worryingly unregulated, and accused of encouraging dangerous sexualities:

> The Internet has become the latest place for promoting the global trafficking and sexual exploitation of women. This global communication network is being used to promote and engage in the buying and selling of women and children. Agents offer catalogues of mail order brides, with girls as young as 13. Commercial sex tours are advertised. Men exchange information on where to find prostitutes and describe how they can be used. After their trips men write reports on how much they paid for women and children and write pornographic descriptions of what they did to those they bought. Videoconferencing is bringing live sex shows to the Internet . . . Global sexual exploitation is on the rise. The profits are high and there are few effective barriers.
>
> (Hughes 2002, 140)

Likewise, for Jeffreys (2008) the international political economy of online adult content is founded on the 'sexual use' of girls and young women, with Internet pornography normalizing the sex industry in the West and legitimating commercial sexual exploitation. Against this, others have pointed to the libertarian and even utopic potential of online sexual materials. For example, Hearns (2006) argues that the Internet acts as a hugely important medium for passing on information on sexuality, sexualized violence and sexual health. This type of content can be especially significant for young 'questioning' groups, especially those considering 'coming out' or needing information on gay and lesbian scenes and communities (which can be daunting for the uninitiated, as Valentine and Skelton 2003 suggest). Message boards and social networking sites like Facebook and Twitter are used by gay and lesbian groups to communicate, make contact with others, organize events, create communities and tell the stories of their lives (Alexander 2003). More widely, Internet technologies can allow for the production of intimacies-at-a-distance (Valentine 2006) in all manner of ways, being tied into practices of coupling, partnership and sex itself.

Sex work advocates and organizations have also found the Internet to be a particular effective medium for disseminating advice and policy recommendations. Client-based sites which provide 'hands-on' reviews of the services provided by sex workers are morally more dubious, but it is significant that some insist on good client behaviour, non-offensive reviews and non-racist and sexist language (Sanders 2008a); notably, information given by clients via the UK-based Punternet website resulted in the arrest of Steven Wright for the murder of five sex workers in Ipswich in 2006. In a somewhat different context, Williams *et al.* (2008) have provided a fascinating analysis of how Singaporean men negotiate new understandings of masculine heterosexual identities through their postings on Sammyboy Times, a commercial sex information site on which men exchange information on their purchases of sex in Batam, Indonesia. Though Williams *et al.* suggest their postings do not wholly transgress offline heteronormative relations, it is clear that in cyberspace, the assumed connections of gender and sexuality become soft, and new forms of masculinity are moulded.

In Chapter Six it was suggested that online adult content now encompasses a diversity of sexual experiences and identities, including pornography made for women by women, porn made by sexual minority groups, porn featuring disabled performers, DIY porn, BDSM videos, fetish scenarios, realcore, chubby sex, public sex, chav porn, covert porn, anime and computer-generated porn, and so on. Irrespective of genre, some of this clearly contains images of non-consensual sex or sexual violence of one kind or another, and Slater (2004, 99) finds much of the material circulating to be 'strikingly organized and policed according to the conventions of off-line mainstream (heterosexual) pornography'. However, Albury (2009) insists that blanket condemnations, such as Catharine MacKinnon's (1994, 1959) assertion that 'pornography in cyberspace is pornography in society, just broader, deeper, worse, and more of it' are way off the mark given the Internet allows people to create new channels of distribution for sexual content outside existing corporate networks. In effect, the Internet provides a new space for people to globally broadcast sexual stories and images that would have previously only had the chance to circulate in very limited circles (Wilkinson 2009).

Irrespective of such arguments concerning the nature of adult content, the rise of Internet pornography has undoubtedly had a massive influence on the visibility and accessibility of sexual materials. Within porn studies there is thus much talk of the 'end of geography' as the 'sticky web' of Internet pornography expands to capture the 'curious clicker' (Johnson 2010). Given the public space of the Internet is characterized by relative anonymity, affordability and accessibility, it allows those in peripheral and poorer regions – especially those outside major metropolitan cores – a way to become active consumers of 'adult content' (Jacobs 2004). This implies that pornography has undergone a shift towards decentralization and heterogeneity as a wider variety of producers and consumers participate in globalized sex markets. For Jacobs (2004), any contemporary study of pornography needs to capture this dispersal, recognizing that the form, content and meaning of pornography in contemporary culture results from 'a network of different factors including (but not limited to) porn performers, producers and distributors, legislation, media, economy, research and various forms of expertise, politics, popular culture and hierarchies of taste' (Paasonen 2009, 587) which can stretch across time and space.

So does this mean cities no longer matter in the production and distribution of circulation? Certainly not, for even if pornography is consumed as a transnational commodity, this does not imply that the geographies of pornography are disorderly or amorphous (Jacobs 2004). Geographic disembedment is illusionary, as the Internet adult industry operates in a space of flows that remains anchored in a space of places. In this sense, it appears some of the arguments deployed in economic geography about the embedded nature of production and the importance of tacit knowledge are relevant to pornography production. Voss (2007) makes such connections when she suggests that in this relatively stigmatized sector, it is extremely difficult to negotiate trust with people who work outside the industry, suggesting that the construction of strong ties within the industry is vital in the creation of new packages of adult entertainment (see also Tibbals 2011). To some extent this explains why the adult industries have tended to be innovators in the use of online and IT technologies: unable to access existing media channels, they have developed their own platforms and media of dissemination to ensure that their product can reach a market. Inter-firm knowledge flows have been vital here, with adult entertainment trade fairs (e.g. Venus Berlin, the Adult Video News Expo held in San Francisco, Erotica LA and Shanghai's Adult Toys Exhibition) being important spaces where reputations and trust are negotiated between individuals in the industry, and where knowledge about the markets for adult consumption is shared (Comella 2008).

As Voss (2007) notes, a key role of such fairs is to encourage *global* copycatting: firms display their latest products, only for other companies to seek to imitate them – something that firms see as inevitable and even beneficial given the lack of patented protection in the sector.

To date there have been few studies of these economic geographies of pornography. The pioneering work of Zook (2003), however, suggests that it might be possible to 'map' the Internet adult industry by specifying the interaction between three sites, namely: the locations where content is produced; the locations where the websites which distribute this content are authored; and the locations where these sites are hosted. In broad terms, the identification of these three key sites corresponds to the vertical integration of production within the pornography industry, from production through to marketing. Typically, some firms focus on production of the underlying adult media, others bundle materials into websites, and still others provide Internet marketing, billing and customer support (Edelman 2009).

The first of these sites relates to the content that adult websites buy to host (the most common practice for pay-to-view providers). Though the creation of content is now relatively easy (in the sense that it merely requires a relatively cheap camcorder and willing participants), the acknowledged centre of porn is San Fernando (or Silicone Valley, as it is sometimes dubbed), which currently accounts for 71 per cent of listed adult entertainment production studios (being home to around 150 companies, including Evil Angel and Vivid). This remarkable agglomeration economy developed, in spite of the efforts of local citizen groups (see Self 2008), by virtue of the business flight of the 1970s that left a glut of low-rent industrial spaces and warehouses where Hollywood wannabe actors and directors could utilize ex-studio equipment and expertise to create the first wave of adult videos (as distinct from *films* – see Simpson 2004). Today, it persists in spite of higher rentals because of the type of factor Voss (2007) identifies, representing a notable cluster in which technical know-how and embodied skills are shared through established networks and inter-firm socialities (significantly, the industry newsletter Adult Video News is based in the Valley).

Emphasizing the continued dominance of the US in studio-produced pornography, Danta (2009) notes that an additional 12 per cent of adult film studios worldwide are located in other parts of Los Angeles; 15 per cent are found in other cities (notably New York, Miami and Las Vegas); and only 2 per cent of studios are located in other countries, principally Russia, the Czech Republic and Hungary (see Case Study 7.3). Edelman (2009) also notes that production is offshoring from San Fernando to Montreal, where tax breaks and immigration laws conspire to create low cost production opportunities. Nevertheless, the geography of porn production remains distinctive in terms of its highly uneven nature (at least in terms of that content sourced from 'professional' studios), with the dominance of US produced pornographic content implicated in a form of 'banal globalization' that valorizes particular body images and sexual practices (see Binnie 2004 on Australian gay porn).

The second location considered by Zook in his mapping of online pornography is the website itself. While these can be fee-paying, by far the largest number are free, sometimes offering limited or low-res content as a bait to encourage the 'curious clicker' to explore. Perhaps paradoxically, much free online content is more hardcore and 'gonzo' than that available through subscription services: paid websites like Penthouse, Hustler and Playboy have relatively soft-core content and make claims to corporate social responsibility so that they can be accessed via as many search engines as possible, and their content downloaded to iPhones and third generation mobiles. Such sites are also massively important for Internet providers given online high res content uses up vast amounts of bandwith, at great cost to the website providers. Mapping domain names of membership websites, Zook (2003, 1274) concludes that the

> UK, Germany, France, and Spain all have a relatively small number of top membership sites compared with their overall presence [on the web] whereas other countries such as the US, Canada, the Netherlands, Denmark, and Australia have a specialization in adult sites.

In absolute terms, the US dominates with around 60 per cent of all paid membership sites, though this figure declined by around 5 per cent between 2001 and 2006 (Zook 2007). One particularly interesting finding here is that a number of adult content sites are registered in locations (such as Antigua, Saint Kitts, the Turks and Caicos Islands) that are otherwise unimportant in domain hosting terms. This is suggestive of an offshoring process that bears comparison to trends in banking for high net worth individuals, perhaps reflecting firms' desire to escape intrusive governmental surveillance.

The third site mapped by Zook is the location of the website itself. In theory, this is the easiest location to map as it can be traced via identification of the IP address of the computer on which the website is hosted. While these could in theory be anywhere in the world, Zook (2007) again suggests that most are hosted in the US, the nation through which the majority of the world's Internet traffic passes. The map of the top 100 Internet websites (measured in terms of Internet traffic, 2009) confirms Zook's observations, and reveals a continuing clustering of IP addresses on the west coast of the

CASE STUDY 7.3

European pornography production: Budapest and Budaporn

Though Hungary was one of the first sources of the silent 'stag' movies that circulated in private hands from the start of the twentieth century, it was in the 1990s that it began to emerge as a capital for European pornographic production. With around 200 full-length films plus many shorter extracts or teasers uploaded to porn websites being produced in the country each year, the country is thought to account for around a quarter of all European pornography. In turn, the industry contributes around 1 per cent of gross domestic product.

The growth of the Hungarian pornography industry is primarily associated with the decline of the Eastern bloc in the late 1980s, subsequent economic reform and, latterly, the country's entry into the European Union (in 2004). All of these have conspired to create low production costs, with government effectively encouraging porn production by allowing the rent of formerly state owned TV production facilities at low cost to foreign porn producers. However, Milter and Slade (2005) argue that it was not only the favourable economic conditions that attracted producers, but also the steady supply of beautiful 'camera-ready' girls. This type of claim rests on assumptions that Hungarian women embody a form of sexuality that is eminently marketable across global marketplaces, associated with a lasciviousness that draws on myths of fiery Roma identity and a 'Magyar' look. At the same time, Hungarian movies became known as rather more hard-core than US productions, with anal sex becoming known as their defining feature.

Yet what is especially interesting about the Hungarian industry is that, unlike most US produced porn, it trades on its placed identity. Budapest, the focus of porn production, takes central stage in many porn movies. Titles like 'From Hungary with love' or 'Buttwoman does Budapest' feature outdoor shooting around the city's photogenic cityscape, with public sex scenarios filmed around the Danube, in the city's parks and even on public trams. As such, 'Budaporn' stands at the fold of East and West, trading on a series of place myths that, in turn, it locates in the midst of a city that it eroticizes.

Further reading: Milter and Slade (2005)

Figure 7.4 Top 100 adult content Internet sites by IP address, 2009 (source: author's data, drawn by Mark Szegner).

US, with other significant hubs in Europe (Amsterdam, Budapest, Paris, London and Brussels): in the Americas beyond the US, only Havana and Buenos Aires feature (see Figure 7.4).

While analyses based on searching for domain names and IP addresses need to be treated with caution given the regional packaging of IP addresses, it is possible here to discern a global geography of porn that remains resolutely routed through particular world cities in the US globalization arena. The evident inertia in the geography of pornography is surprising given the rapid technological changes in the sector as well as the low start up costs associated with pornographic production and uploading. However, the shifts that Zook (2007) discerns, in terms of a gradual movement away from US cities towards European cities, suggest there are specific 'pull' factors diffusing the adult entertainment industry into world cities in other globalization arenas. Bringing the type of analysis Zook offers into dialogue with studies of the places – and cities – where Internet pornography is consumed is thus important if we are to better understand the world city networks of pornography. What is of course missing here is reliable data on the real flows of Internet traffic that constitute this form of 'globalization from below': simply put, we need to know much more about who is viewing what, when and where before we can truly map the spaces of adult content (see also Juffer 1998; Lumby *et al.* 2008; Edelman 2009).

Conclusion

To date, world city research has said little about matters of sexuality, despite the evidential importance of sex for the economies of world cities and, conversely, the importance of world cities for the globalized sex industry. This chapter has offered a broad overview of some of the ways that world cities research might usefully engage with themes of sex and sexuality, arguing that the increased possibilities for migration and mobility associated with globalization mean that contemporary cities are, perhaps more than ever, contact zones where people of different social and cultural backgrounds mix. In such cities, sex and bodies hence become 'commodities that can be packaged, advertised, displayed, and sold on a global scale' (Wonders and Michalowski 2001, 117) with the polarization between a high wage service class and a low wage servicing class creating both supply and demand for sexual services (Sassen 2002). This given, the prevalence of commercial sex at the heart of the major 'decision making centres' of the global economy is unsurprising given their status as spaces in which migrants constitute a significant share of the workforce, whether as members of the transnational creative class or as the workers who populate the brothels, bars and nightclubs which act as crucial sites of sexual–economic exchange.

Arguing that the sexual economy is not always a sideshow to the 'real work' of world cities, this chapter has also suggested that some cities need to be considered as 'erotic cities' – de facto centres of sex – whose sexual economies are integral to their global reach and influence. These are not just spaces where sex is consumed, but sites where sexual markets are developed, global networks of sex-related migration set in motion and adult entertainment produced. The importance of certain cities in the articulation of a global pornographic industry is a case in point, with this multi-billion dollar industry rooted in a relatively limited range of cities that display certain competitive and comparative advantages in the production and distribution of pornography.

The diverse embodied and sexualized economies characteristic of world cities suggest that metropolitan centres remain the hubs of a global sexual economy that trades in the production and consumption of desire. In this context, the rise of Internet pornography and virtualized sex work does little to challenge the idea that the city is the paradigmatic site of sexual encounter (Houlbrook 2006). Real and virtual sex are inseparable, intimately entwined in those world cities that orchestrate the global economy. Global lines of desire are thus inescapably mapped onto, and out of, specific metropolitan landscapes – for example, the beaches of Dubai, nightclubs in New York,

Cape Town's gay village, the streets of Mumbai, the casinos of Las Vegas or Budapest's picturesque city centre – reflecting uneven geographies of globalization, heteronormativity and capital accumulation.

But even if we recognize that there is a cadre of world cities that plays a disproportionate role in the reproduction of a global sexual economy, we need to remain wary of the idea that there is a unified global sexual culture or, equally, the idea that the flows involved here are unidirectional, from metropolitan West to the non-West. For, as Judith Halberstam (2005) proposes in her critique of what she terms the 'metronormativity' of queer studies, every city represents a unique re-articulation of the heterosexual and homosexual binary, inflected through specific modalities of class, ethnicity and location (see also Oswin 2006; Tucker 2009). This point is important both for geographies of sexuality as well as queer political projects, for it challenges the emphasis on Western cities as the most important centres in the creation of sexual knowledge. Indeed, Halberstam (2005, 38) argues it is vital to consider non-urban sexualities in relation to 'the dominant metropolitan model of gay male sexual exchange', as well as an urgent need for a trans-regional analysis of sexuality that does not privilege the male, cosmopolitan, elite (see also Brown 2008b). Rather than remaining fixated on what happens within relatively small areas of Western, metropolitan centres, such as gay villages or red light districts, it appears that analyses of urban sexuality must employ a more expansive geographical imagination in which local and global are always entwined, and where urban sexualities are not just enacted by those in metropolitan centres.

Further reading

Altman (2002) considers a broad range of debates around the internationalization of sex, with particular emphasis on the neoliberal politics of HIV/AIDS and the related development of the 'global gay'.

Binnie (2004) provides a geographer's perspective on the socialities and sexualities developed at a distance, and devotes particular attention to global queer networks.

Manderson and Jolly (1997) provide a collection of empirical examples that trace the impacts of some of the processes outlined in this chapter in Pacific Asia, with considerable attention devoted to the dynamics of trafficking, prostitution and sex tourism.

8 Conclusion

Learning objectives

- To be able to draw conclusions about the importance of the city in structuring our sexual lives – and vice versa.
- To begin to understand how cities might be re-imagined and re-worked to accommodate different forms of sexual and intimate life.

This book began by stressing the relative neglect of sexuality in urban studies. Given that sexuality appears so central to our lives, and is endlessly debated, it was argued that it is impossible to make sense of the city without considering its sexual dimensions. To demonstrate this, the book has spiralled off in numerous different directions, considering the way that urban space is experienced and felt differently depending on one's sexual orientation, as well as noting the importance of the city in structuring, directing and giving meaning to people's sexual lives. As has been demonstrated, whether we identify as lesbian, straight, gay or bisexual (or perhaps none of those categories), sexuality has a profound impact on our experience of the city given urban space can open up, or close down, different spaces for sexual expression. Through both the regulation enacted by the state and law, and the self-regulation associated with the desire to be desired, the city is hence divided into spaces that are sexualized in different ways. The boundedness of these spaces is often clear, marked out through borders that are legally defined and defended: as examples throughout this book have demonstrated, the isolation and enclosure of 'deviant' and 'dangerous' sexualities through different techniques of state-sponsored governmentality has been an important tactic used in the ordering of the city. Concurrently, the state promotes other forms of sexuality that are thought to be beneficial to the social and economic functioning of the city (whether through the promotion of 'gay villages' as consumer spaces, the licensing of lap-dancing clubs or the sponsorship of family housing).

However, the emphasis in much of this book on the control of sexuality and space should not distract from the city's *polymorphous perversity*, with the ultimate impossibility of fixing sexualities in space evident in examples of performances that tactically transform, disturb or transgress expectations of where different forms of sex, intimacy and love should take place. The use of the city's green spaces, public toilets, car parks and derelict sites for anonymous sexual encounter is one such illustration of how the city's spaces can be sexualized through performances that temporarily invert dominant orders of public–private. Throughout this book, other examples have shown that any number of sexualized minorities – sex workers, BDSM practitioners, naturists – have mobilized across the public and private spaces of the city, using the former to argue that they should be allowed to exercise sexual self-autonomy in the latter. In some instances, repeated performances of this kind have also served to normalize lesbian and gay sexualities and forced an acceptance of their legality, with the emergence of visible gay villages in the urban West providing a notable example of how once highly marginalized sexualities can be normalized in the landscape (Knopp 1992; Forest 1995). Likewise, adult businesses are also becoming highly visible in many cities, suggesting a grudging acceptance of recreational and commercial sex as part of the urban scene.

Given the visibility of such new sites of sexual identification in Western cities, it is tempting to argue that the city is becoming 'softer' as traditional sexual moralities and norms are challenged. The seeming diminution of the traditional family, the increased rates of divorce and the legal recognition of same-sex civil unions all suggest something fundamental has happened to sexuality in recent decades, with the idealization of the nuclear family being superseded by a wider range of possible household types and lifestyle choices. In its 'plastic' incarnation, sex has become recreational, with the trade in sperm and human embryos, in vitro fertilization and reproductive technologies meaning that sex itself is no longer even required for procreation. The implication is, seemingly, that people are more able to choose sexual lifestyles and identities beyond the taken-for-granted norm of the married, co-resident, heterosexual couple with children. Giddens' (1992) identification of the dominance of 'confluent' love based on mutual satisfaction rather than life-long commitment highlights this putative shift in sexual ethics, something registered in the increased purchase of sexual services (Bernstein 2007), the consumption of pornography (Smith 2007b), the use of online dating sites (Couch and Liamputtong 2008) and pursuit of hetero-flexible lifestyles

(Wosick-Correa and Joseph 2008). At the same time, the civil rights and protections now granted to lesbian and gay identified individuals in many nations suggests that they enjoy the same spatial freedoms and rights to the city as heterosexually identified individuals – hence the title of Jeffrey Weeks' (2007) review of profound revolution in British intimate life, *The World We Have Won*.

But this sexual 'diversification and dispersion' (Sigusch 1998) remains geographically uneven. As has been described, not all cities are equally open to sexual diversity – and many of the observations in this book have been limited to large, Western cities. And even in the urban West, the sexuality normalized in the city remains restricted: despite the widely noted transformation of intimacy, there are still clear limits to sexual citizenship. 'Dangerous' sexualities still abound, lurking in the corners of the city, on the margins and in the unregulated spaces of the Internet. Sexualities can be accommodated within the contemporary city, it appears, only if they can be described as consensual. This means that the contemporary Western city is best described not as purely heteronormative, nor homonormative, but as reproducing certain assumptions about the importance of an emotionally mature sexuality that takes as its object of desire a *consenting* adult partner (and certainly not a coerced or exploited person, a child or an animal, nor anyone whose consent is not explicable to the state and law). Singleness also remains suspect, whether people live alone by choice or appear unable to form relationships (Cobb 2007).

The Western city therefore appears to be vital in the production of new norms of intimacy, with the norm of heterosex in marriage appearing to have been replaced by a privileging of consensual sex within the (assumed to be emotionally fulfilling) couple form. Underpinning all this, of course, is the urban economy. Sex sells, with the commercialization of sexuality appearing inevitable as capitalism fragments intimacy, romance and sex into as many saleable pieces as possible to maximize profit. Indeed, Bauman (2003) relates the transformation of intimacy to the rise of the consumer society – a society where the marketplace appears to offer the security and happiness unobtainable elsewhere. Of course, consumption never assuages lack, and merely bequeaths further acts of consumption, condemning us to live our lives as a series of shopping trips. In this sense, sex is just another form of consumption, often quick and disposable.

Whether in its more disposable or durable forms, the commodification of sex and relationships can be highly profitable, as the recent growth of the global pornography, sex retailing and sexuopharmaceutical industries testifies (witness the global reach of brands like Viagra, Hustler, Durex or Spearmint Rhino). It is here that contradictions emerge, as the traditionally taboo status of commercial sex has been vital in ensuring it has value in the marketplace. Making sex increasingly accessible and visible risks it becoming ubiquitous, and while the sex industries are sometimes described as recession-proof, it appears that if access to sexual materials becomes too easy, or free, then there is a need to introduce new controls and prohibitions to limit their visibility. For example, the emergence of a new DIY economy in online pornography, facilitated by new interactive online platforms, allows individuals to make money out of their own sexually explicit material, or even to give it away for free (Paasonen 2010). This 'democratisation of desire', and the related promotion of sexual freedom, is one that has been both exploited and curtailed by the mainstream pornography industry, which co-opts such content through the construction of elaborate affiliate programmes (Johnson 2010) while simultaneously encouraging anti-piracy and censorship regulation of the Internet on the basis that too much sexual freedom encourages child pornography and sexual exploitation (Maddison 2010).

The continuing regulation of sexual commerce, and the state's insistence that we are good, self-governing consumers of sex, is explicable in the context of contemporary neoliberal discourses which encourage the combination of informed consent, contractual interaction, free market choice and responsibility. Distinctions between good, neoliberal subjects and 'non-citizens' who refuse to take responsibility for their own lives often remain based on understandings of sexual practices – for example, irresponsible buying of sex from migrant women (Brooks-Gordon 2010) or knowingly engaging in unsafe sex (Tomso 2008) – perpetuating a 'restricted' urban economy in which desire is always channelled to corporate, capitalist ends. The city plays an active role in perpetuating this economy via the spatial identification and isolation of spaces of 'sexual risk'. In this sense, the licensing or granting of planning permission for brothels (Brents and Hausbeck 2007), lap-dancing clubs (Hubbard *et al.* 2009), gay saunas (Prior 2008) or sex shops (Coulmont and Hubbard 2010) might all be read as symbolic of the liberalization of sex and the city, but each remains an attempt to survey, contain and 'fix' particular illiberal subjects.

So are we seeing the emergence of more sexually liberal and less repressed cities? Perhaps, but it appears naive to talk of an urban sexscape which is open to all. The continuing circulation of discourses of sexual danger and moral threat ensures that many sites of sexual difference remaining fiercely embattled (Jackson 2009). So even if the city now seems to offer a diversity of spaces for the celebration of sexual difference, there are clearly limits to the difference that can be accommodated. This has been particularly noted in debates concerning the queering of cities, with clearly defined gay consumption districts in cities including Toronto (Nash 2006), Sydney (Ruting 2007) and Manchester (Bell and Binnie 2004) described as spaces which effectively exclude many queer identified people on the basis of their age, ethnicity, class or sexual predilections. Maddison captures this when reflecting on his own experience of such districts:

> In the bars in Soho's Compton Street, or along Manchester's Canal Street, the small town boy may find a cosmopolitan culture welcoming in its beauty, availability, glamour and gayness, but he will have to manage his complex emotional needs, his pent-up distresses, frustrations, his curiosities and passions into modes communicable through cruising rituals and performances of bodily commodification . . . There was a time when such small town boys could have found, in less salubrious bars, the kinds of shockingly unfashionable but infinitely valuable old Queens who inducted generations of us, into ways of resisting, flouting, bitching, screaming and caring that we didn't need a Platinum card for, but which I for one couldn't have lived without. You can't sell that kind of wisdom; you can't professionalise it, and its doyennes don't connote the kind of endless youthfulness that (post)modern gay venues require.
>
> (Maddison 2002, 155)

Nash (2010) likewise argues that 'queer' spaces (such as gay villages) can be essentially disciplinary, fixing gay identities in ways that are distinctly un-queer. The experiences reported by many gender-indeterminate and transgendered individuals in gay villages underline that these spaces are not always welcoming of difference (something Browne and Lim 2010 note in their exploration of trans lives in the 'gay capital of the UK', Brighton). Doan (2004, 68) similarly alleges there is 'little to no visible gender queerness or any indication that such variance is tolerated' in such sexually liberatory spaces, and that 'even in San Francisco's Castro District, often considered the archetype of queer space, the streets are full of well muscled men and even the window displays are masculinely gendered'. Nast (2002) also describes the 'phallic rimming' of Chicago's 'Boy's Town' through strategies of signage that make it clear this is a gay *male* space that is far from radically inclusive.

Hence, while early studies of 'gay spaces' emphasized their edgy and resistive elements, later literature has argued that such spaces have been systematically 'de-queered' as they become colonized by the market (Visser 2008; Weiss 2008; Brown and Browne 2009; Doan and Higgins 2011). This leads to a stark conclusion: cities may be getting sexier, but only *profitable* expressions, representations and performances of sexuality are being encouraged. This book is hence one that shows that sexuality matters in city life not just because it is a question of personal choice that matters to the individual, but because it is deeply implicated in the ways that cities *work* as sites of social and economic reproduction.

Bibliography

Aalbers, M. B. (2005). "Big Sister is watching you! – gender interaction and the unwritten rules of the Amsterdam red-light district." *Journal of Sex Research* **42**(1): 54–62.

Abbott, E. (2000). *A History of Celibacy*. New York, Scribner.

Abel, G. M., Fitzgerald, L. J. and Brunton, C. (2009). "The impact of decriminalisation on the number of sex workers in New Zealand." *Journal of Social Policy* **38**: 515–31.

Abraham, J. (2009). *Metropolitan Lovers: the homosexuality of cities*. Minneapolis, University of Minnesota Press.

Acker, M. (2009). "Breast is best . . . but not everywhere: ambivalent sexism and attitudes toward private and public breastfeeding." *Sex Roles* **61**(7): 476–90.

Acton, W. (1870). *Prostitution Considered in its Moral, Social and Sanitary Aspects in London and Other Large Cities and Garrison Towns with Proposals for the Control and Prevention of its Attendant Evils*. London, John Churchill.

Adler, S. and Brenner, J. (1992). "Gender and space: lesbians and gay men in the city." *International Journal of Urban and Regional Research* **16**: 24–34.

Aggleton, P. (ed.) (1999). *Men who Sell Sex: international perspectives on male prostitution and HIV/AIDS*. Philadelphia, Temple University Press.

Agustin, L. (2005). "The cultural study of commercial sex." *Sexualities* **8**(5): 618–31.

—— (2007). *Sex at the Margins: migration, labour markets and the rescue industry*. London, Zed Publishing.

Aitken, S. (1999). *Family Fantasies and Community Space*. New Jersey, Rutgers University Press.

—— (2009). *The Awkward Spaces of Fathering*. Chichester, Ashgate.

Alaimo, S. (2010). "The naked word: The trans-corporeal ethics of the protesting body." *Women & Performance: a journal of feminist theory* **20**(1): 15–36.

Albert, A., and Warner, D. (1995). "Condom use among female commercial sex workers in Nevada's legal brothels." *American Journal of Public Health* **85**(11): 1514.

Albury, K. (2002). *Yes means Yes: getting explicit about heterosex*. London, Unwin Hyman.

—— (2009). "Reading porn reparatively." *Sexualities* **12**(5): 647–53.

Aldrich, R. (2004). "Homosexuality and the city: an historical overview." *Urban Studies* **41**: 1719–37.

Alexander, J. (2003). "Queer webs: representations of LGBT people and communities on the world wide web." *International Journal of Sexuality and Gender Studies* **7**(2/3): 77–84.

Alexander, M., MacLaren, A. C. and O'Gorman, K. D. (2010). "Beyond the brothel: hospitable spaces for sex in the 21st century." In M. Alexander, A. C. MacLaran and K. D. O'Gorman (eds) *Commercial Hospitality: exploring contexts and boundaries*. Oxford, Goodfellow.

Allison, A. (1994). *Nightwork: sexuality, pleasure and the performance of masculinity in a Tokyo hostess club*. Chicago, University of Chicago Press.

Altman, D. (2002). *Global Sex*. Chicago, University of Chicago Press.

—— (2004). "Sexuality and globalization." *Sexuality Research and Social Policy* **1**(1): 63–8.

Anderson, E. (1989). "Sex codes and family life among poor inner-city youths." *The Annals of the American Academy of Political and Social Science* **501**: 59–78.

Andersson, J. (2011). "Vauxhall's postindustrial pleasure gardens: 'death wish' and hedonism in 21st century London." *Urban Studies* **48**(1): 85–100.

Andriotis, K. (2010). "Heterotopic erotic oases: the public nude beach experience." *Annals of Tourism Research* **37**(4): 1076–96.

Ashford, C. (2008). "Sexuality, public space and the criminal law: the cottaging phenomenon." *Journal of Criminal Law* **71**(6): 506–18.

—— (2010). "Socio-legal perspectives on gender, sexuality and law: editorial." *Liverpool Law Review* **31**(1): 1–12.

Ashworth, G., White, P. and Winchester, H. P. M. (1988). "The red-light district in the west European city: a neglected aspect of the urban landscape." *Geoforum* **1**(2): 201–12.

Askew, M. (1998). "City of women, city of foreign men: working spaces and re-working identities among female sex workers in Bangkok's tourist zone." *Singapore Journal of Tropical Geography* **19**(2): 130–50.

Asthana, A. (2007). "Amsterdam closes a window on its red-light tourist trade." *The Observer*, 23 September: www.guardian.co.uk/travel/2007/sep/23/travelnews.amsterdam (accessed 3 August 2011).

Atkins, G. (2003) *Gay Seattle: stories of exile and belonging*. Seattle, University of Washington Press.

Attfield, J. (2007). *Bringing Modernity Home: writings on popular design and material culture*. Manchester, Manchester University Press.

Attwood, F. (2005). "Fashion and passion: marketing sex to women." *Sexualities* **8**(4): 392–406.

Attwood, F. and Smith, C. (2010). "Extreme concern: regulating 'dangerous pictures' in the United Kingdom." *Journal of Law & Society* **37**(1): 171–88.

Babb, F. E. (2007). "The intimate economies of Bangkok: tomboys, tycoons, and Avon ladies in the global city." *GLQ: a journal of lesbian and gay studies* **13**(1): 111–23.

Bakhtin, M. (1984). *Rabelais and his World*. Gary, Indiana University Press.

Barber, A. (2010). *From the Red Light District to the Red Carpet: the gentrification of Amsterdam's red light district and its impact on the democratic citizenship rights of sex workers*. Canadian Political Science Association Papers. Concordia: www.cpsa-acsp.ca/papers-2010/Barber.pdf (accessed 6 May 2011).

Barcan, R. (2001). "'The moral bath of bodily unconsciousness': female nudism, bodily exposure and the gaze." *Continuum: Journal of Media & Cultural Studies* **15**(3): 303–17.

—— (2004). *Nudity: a cultural anatomy*. Oxford, Berg.

Barker, M. and Langdridge, D. (eds) (2010). *Understanding Non-Monogamies*. London, Routledge.

Barnes, J., Dukes, T., Tewksbury, R. and Troye, T. M. (2009). "Analyzing the impact of a statewide residence restriction law on South Carolina sex offenders." *Criminal Justice Policy Review* **20**(1): 21.

Barron, D. L. (1999). "Sex and single girls in the twentieth century city." *Journal of Urban History* **25:** 838–47.

Bartley, P. (2000). *Prostitution: prevention and reform in England, 1860–1914*. London, Routledge.

Basil, M. (2008). "Japanese love hotels: protecting privacy for private encounters." *European Advances in Consumer Research* **8**: 508–14.

Bataille, G. (1970). *L'Érotisme*. Paris, Éditions de Minuit.

Bauman, Z. (2003). *Liquid Love: on the frailty of human bonds*. Oxford, Polity.

BBC (2003). "Stag parties ruin Blackpool trade." Online. Available http://news.bbc.co.uk/1/hi/england/lancashire/3093043.stm (accessed 3 August 2011).

Beasley, C. (2008). "Rethinking hegemonic masculinity in a globalizing world." *Men and Masculinities* **11**(1): 86.

Bech, H. (1997). *When Men Meet: homosexuality and modernity*. Chicago, Chicago University Press.

Beck, U. and Beck-Gernsheim, E. (1995). *The Normal Chaos of Love*. Oxford, Polity.

Beemyn, B. (1997). *Creating a Place for Ourselves: lesbian, gay, and bisexual community histories*. London, Routledge.

Bell, D. (1995). "Pleasure and danger: the paradoxical spaces of sexual citizenship." *Political Geography* **14**(2): 139–53.

—— (2001). "Fragments for a queer city." In D. Bell, J. Binnie, R. Holliday, R. Longhurst and R. Peace (eds) *Pleasure Zones: bodies, cities, spaces*. Syracuse, Syracuse University Press.

—— (2008). "Destination drinking: toward a research agenda on alcotourism." *Drugs: education, prevention, and policy* **15**(3): 291–304.

—— (2009a). "Bodies, technologies, spaces: on 'dogging'." *Sexualities* **9**(4): 387–407.

—— (2009b). "Surveillance is sexy." *Surveillance & Society* **6**(3): 203–12.

Bell, D. and Binnie, J. (1998). "Theatres of cruelty, rivers of desire." In N. Fyfe (ed.) *Images of the Street: planning, identity, and control in public space*. London, Routledge.

—— (2004). "Authenticating queer space: citizenship, urbanism and governance." *Urban Studies* **41**(9): 1807–20.

Bell, D. and Holliday, R. (2000). "Naked as nature intended?" *Body and Society* **6**(3–4): 127–40.

Bell, D. and Valentine, G. (eds) (1995a). *Mapping Desire: geographies of sexualities*. London, Routledge.

Bell, D. and Valentine, G. (1995b). "Queer country: rural lesbian and gay lives." *Journal of Rural Studies* **11**(2): 113–22.

Bell, D., Binnie, J., Cream, J. and Valentine, G. (1994). "All hyped up and no place to go." *Gender, Place & Culture* **1**(1): 31–47.

Bell, S. (1994). *Reading, Writing, and Rewriting the Prostitute Body*. Gary, Indiana University Press.

Bellis, M., Hughes, K., Thomsen, R. and Bennett, A. (2004). "Sexual behaviour of young people in international tourist resorts." *Sexually Transmitted Infections* **80**(1): 43.

Bellis, M. A., Watson, F. L. D., Hughes, S., Cook, P. A., Downing, J., Clark, P. and Thornson, R. (2007). "Comparative views of the public, sex workers, businesses and residents on establishing managed zones for prostitution: analysis of a consultation in Liverpool." *Health & Place* **13**(3): 603–16.

Berlant, L. (1997). *The Queen of America goes to Washington City: essays on sex and citizenship*. Durham, Duke University Press.

Berlant, L. and Freeman, E. (1992). "Queer nationality." *Boundary* **19**(1): 149–80.

Berlant, L. and Warner, M. (1998). "Sex in public." *Critical Inquiry* **24**(2): 547–66.

Berman, J. (2003). "(Un)popular strangers and crises (un)bounded: discourses of sex-trafficking, the European political community and the panicked state of the modern state." *European Journal of International Relations* **9**(1): 37–86.

Berman, M. (1983). *All that is Solid Melts into Air: the experience of modernity*. New York, Verso.

—— (1986). "Take it to the streets." *Dissent* (fall): 490–6.

—— (2001). "Too much is not enough: metamorphoses of Times Square." In T. Smith (ed.) *Impossible Presence: surface and screen in the photogenic era*. Chicago, University of Chicago Press.

—— (2006). *On the Town: one hundred years of spectacle in Times Square*. New York, Random House.

Bernstein, E. (2007). *Temporarily Yours: intimacy, authenticity, and the commerce of sex*. Chicago, University of Chicago Press.

Berube, M. (2003). "The history of gay bathhouses." In W. Woods and D. Binson (eds) *Gay Bathhouses and Public Health Policy*. Boston, Harrington Park Press.

Besio, K., Johnston, L. and Longhurst, R. (2008). "Sexy beasts and devoted mums: narrating nature through dolphin tourism." *Environment and Planning A* **40**(5): 1219–34.

Bhattacharyya, G. (2002). *Sexuality and Society: an introduction*. London, Routledge.

Bieri, S. and Gerodetti, N. (2007). "'Falling women'–'saving angels': spaces of contested mobility and the production of gender and sexualities within early twentieth-century train stations." *Social & Cultural Geography* **8**(2): 217–34.

Binnie, J. (2004). *The Globalization of Sexuality*. London, Sage.

Binnie, J. and Skeggs, B. (2004). "Cosmopolitan knowledge and the production and consumption of sexualised space: Manchester's gay village." *The Sociological Review* **52**(1): 39–61.

Binnie, J. and Valentine, G. (1999). "Geographies of sexuality – a review of progress." *Progress in Human Geography* **23**(2): 175–87.

Binnie, J., Holloway, J., Millington, S. and Young, C. (2006). "Conclusion: the paradoxes of cosmopolitan urbanism." In J. Binnie, J. Holloway, S. Millington and C. Young (eds) *Cosmopolitan Urbanism*. London, Routledge.

Bird, R. (2009). "'Sex, sun, soccer': stakeholder-opinions on the sex industry in Cape Town in anticipation of the 2010 FIFA Soccer World Cup." *Urban Forum* **20**: 33–46.

Birken, L. (1988). *Consuming Desire: sexual science and the emergence of a culture of abundance, 1871–1914*. Ithaca, Cornell University Press.

Bishop, R. and Robinson, J. S. (1998). *Night Market: sexual cultures and the Thai economic miracle*. New York, Routledge.

Blidon, M. (2008). "La casuistique du baiser." *EchoGeo* **5** (June/August): http://echogeo.revues.org/5383 (accessed 6 May 2011).

Blomley, N. (2004). *Unsettling the City: urban land and the politics of property*. New York, Routledge.

—— (2005). "The borrowed view: privacy, propriety, and the entanglements of property." *Law and Social Inquiry: journal of the American Bar Foundation* **30**(4): 617–61.

Blunt, A. and Downing, R. (2006). *Home*. London, Routledge.

Bondi, L. (1997). "Sex in the city." In R. Fincher and J. Jacobs (eds) *Cities of Difference*. New York, Guilford Press.

Booth, C., Darke, J. and Yeandle, S. (1996). *Changing Places: women's lives in the city*. London, Paul Chapman.

Booth, D. (1997). "Nudes in sand and perverts in the dunes." *Journal of Australian Studies* **53**(2): 148–60.

Boso, L. (2009). "(Trans)gender-inclusive equal protection: analysis of public female toplessness." *Law & Sexuality: rev. lesbian, gay, bisexual & transgender legal issues* **18**: 143–9.

Bouthillette, A. (1994). "Gentrification by gay male communities: a case study of Toronto's Cabbage Town." In S. Whittle (ed.) *The Margins of the City: gay men's urban lives*. Aldershot: Arena.

Boyd, J. (2010). "Producing Vancouver's (hetero) normative nightscape." *Gender, Place & Culture* **17**(2): 169–89.

Boydell, S., Searle, G. and Small, G. (2007). *The Contemporary Commons: understanding competing property rights*. State of Australian Cities Conference (SOAC 2007) – Growth, Sustainability and Vulnerability of Urban Australia. Adelaide, South Australia, SOAC/UniSA.

Boyer, C. (2001). "Twice-told stories: the double erasure of Times Square." In I. Borden, A. Kerr, J. Rendell and B. Pivaro (eds) *The Unknown City*. Boston, MIT Press.

Brader, C. (2005). "A world on wings: young female workers and cinema in World War I." *Women's History Review* **14**(1): 99–118.

Braidotti, R. (1994). *Nomadic Subjects: embodiment and sexual difference in contemporary feminist theory*. New York, Columbia University Press.

Brennan, D. (2010). "Sex tourism and sex workers' aspirations." In R. Weitzer (ed.) *Sex for Sale: prostitution, pornography and the sex industry*. New York, Routledge.

Brents, B. and Hausbeck, K. (2005). "'Violence and legalized brothel prostitution in Nevada: examining safety, risk and prostitution policy." *Journal of Interpersonal Violence* **20**(3): 270–95.

—— (2007). "Marketing sex: US legal brothels and late capitalist consumption." *Sexualities* **10**(4): 425–39.

—— (2010). "Sex work now: what the blurring of boundaries around the sex industry means for sex work, research and activism." In M. H. Ditmore, A. Levy and A. Willman (eds) *Sex Work Now*. New York, Zed Books.

Brents, B. and Sanders, T. (2010). "Mainstreaming the sex industry: economic inclusion and social ambivalence." *Journal of Law & Society* **37**(1): 40–60.

Brents, B., Jackson, C. A. and Hausbeck, K. (2009). *The State of Sex: tourism, sex and sin in the new American heartland*. New York, Routledge.

Brewis, J. and Linstead, S. (2000). *Sex, Work and Sex Work: eroticising organizations*. London, Routledge.

Brickell, C. (2000). "Heroes and Invaders: gay and lesbian pride parades and the public/private distinction in New Zealand media accounts." *Gender, Place and Culture* **7**(2): 163–78.

Bromley, R., Thomas, C. and Millie, A. (2000). "Exploring safety concerns in the night-time city: revitalising the evening economy." *The Town Planning Review* **71**(1): 71–96.

Brooks-Gordon, B. (2010). "Bellwether citizens: the regulation of male clients of sex workers." *Journal of Law & Society* **37**(1): 145–70.

Brown, D., Farrier, D. and Weisbrot, D. (2006). *Criminal Laws: materials and commentary on criminal law and process*. Sydney, The Federation Press.

Brown, G. (2007). "Mutinous eruptions: autonomous spaces of radical queer activism." *Environment and Planning A* **39**(11): 2685–98.

—— (2008a). "Ceramics, clothing and other bodies: affective geographies of homo-erotic cruising encounters." *Social & Cultural Geography* **9**(8): 915–32.

—— (2008b). "Urban (homo)sexualities: ordinary cities and ordinary sexualities." *Geography Compass* **2**(4): 1215–31.

Brown, G. and Browne, K. (2009). "Gay space." In R. Hutchinson (ed.) *Encyclopedia of Urban Studies*. Newbury Park, Sage.

Brown, G., Lim, J. and Browne, K. (2007). "Introduction, or why have a book on geographies of sexualities?" In K. Browne, J. Lim and G. Brown (eds) *Geographies of Sexualities*. London, Ashgate.

Brown, G., Browne, K., Elmhirst, R. and Hutta, S. (2010). "Sexualities in/of the Global South." *Geography Compass* **4**(10): 1567–79.

Brown, M. (2000). *Closet space: geographies of metaphor from the body to the globe*. London, Routledge.
—— (2008). "Urban geography plenary lecture: public health as urban politics, urban geography: venereal biopower in Seattle, 1943–83." *Urban Geography* **30**: 1–29.
Brown, M. and Knopp, L. (2008). "Queering the map: the productive tensions of colliding epistemologies." *Annals of the Association of American Geographers* **98**(1): 40–58.
—— (2010). "Between anatamo- and bio-politics: geographies of sexual health in wartime Seattle." *Political Geography* **29**(3): 392–403.
Brown, M. and Rasmussen, C. (2009). "Bestiality and the queering of the human animal." *Environment and Planning D: society and space* **28**: 99–118.
Browne, K. (2007). "A party with politics? (Re)making LGBTQ Pride spaces in Dublin and Brighton." *Social & Cultural Geography* **8**(1): 63–87.
Browne, K. and Lim, J. (2010). "Trans lives in the 'gay capital of the UK'." *Gender, Place & Culture* **17**(5): 615–33.
Browne, K., Lim, J. and Brown, G. (eds) (2007). *Geographies of Sexualities*. London, Ashgate.
Browne, K., Cull, M. and Hubbard, P. (2010). 'The diverse vulnerabilities of lesbian, gay, and trans sex workers in the UK.' In K. Hardy, S. Kingston and T. Sanders (eds) *New Sociologies of Sex Work*. London, Ashgate.
Brownlow, A. (2009). "Keeping up appearances: profiting from patriarchy in the nation's 'safest city'." *Urban Studies* **46**(8): 1680–701.
Buck-Morss, S. (1991). *The Dialectics of Seeing: Walter Benjamin and the Arcades Project*. Cambridge, MA, MIT Press.
Budgeon, S. (2008). "Couple culture and the production of singleness." *Sexualities* **11**(3): 301–25.
Bulkens, M. (2009). "'A delicious leisure activity?' Spatial resistance to hetero-normativity in public spaces." Unpublished MSc thesis, Wagenningen University.
Butler, J. (1993). *Bodies that Matter: on the discursive limits of "sex"*. London, Routledge.
—— (1999). *Gender Trouble: feminism and the subversion of identity*. London, Routledge.
Cabezas, A. L. (2004). "Between love and money: sex, tourism and citizenship in Cuba and the Dominican Republic." *Signs* **29**(4): 987–1015.
Califia, P. (1994). *Public Sex: the culture of radical sex*. New York, Cleis Press.
Cameron, S. (2004). "Space, risk and opportunity: the evolution of paid sex markets." *Urban Studies* **41**(9): 1643–57.
Carline, A. (2010). "Ethics and vulnerability in street prostitution: an argument in favour of managed zones." *Crimes and Misdemeanours: deviance and the law in historical perspective* **3**(1): 20–53.
—— (2011). "Constructing the subject of prostitution: a Butlerian reading of the regulation of sex work." *International Journal of the Semiotics of Law* **24**: 61–78.

Carter, D. (2005). *Stonewall: riots that sparked a revolution*. London, St Griffins Press.

Casey, M. (2004). "De-dyking queer space(s): heterosexual female visibility in gay and lesbian spaces." *Sexualities* **7**(4): 446–60.

Caslin, S. (2010). "Flappers, amateurs and professionals: the spectrum of promiscuity in 1920s Britain." In K. Hardy, S. Kingston and T. Sanders (eds) *New Sociologies of Sex Work*. London, Ashgate.

Castells, M. (1983). *The City and the Grassroots*. Berkeley, University of California Press.

Catungal, J. and McCann, E. (2010). "Governing sexuality and park space: acts of regulation in Vancouver, BC." *Social & Cultural Geography* **11**(1): 75–94.

Chalfen, R. (2009). "'It's only a picture': sexting, smutty snapshots and felony charges." *Visual Studies* **24**(3): 258–68.

Chaplin, S. (2007). *Japanese Love Hotels: a cultural history*. London, Routledge.

Chatterton, P (2002). "Governing nightlife: profit, fun and (dis)order in the contemporary city." *Entertainment Law* **1**(2) Summer: 23–49.

Chatterton, P. and Hollands, R. (2003). *Urban Nightscapes: youth cultures, pleasure spaces and corporate power*. London, Routledge.

Chauncey, G. (1995). *Gay New York: gender, urban culture and the making of the gay male world 1890–1940*. New York, Basic Books.

Chelsea News and Advertiser (1871). Cited in P. Howell (1995). "Victorian sexuality and the moralisation of Cremorne Gardens." In J. Sharp, P. Routledge, C. Philo and R. Paddison (eds), *Entanglement of Power: geographies of domination/resistance*. London: Routledge.

Chisholm, D. (1999). "The traffic in free love and other crises: space, pace, sex and shock in the city of late modernity." *Parallax* **5**(3): 69–89.

—— (2005). *Queer Constellations: subcultural space in the wake of the city*. Minneapolis, University of Minnesota Press.

Churchill, D. S. (2004). "Mother Goose's map: tabloid geographies and gay male experience in 1950s Toronto." *Journal of Urban History* **30**(6): 826–52.

Clark, A. (2008). *Desire: a history of European sexuality*. London, Routledge.

Cobb, M. (2007). "Lonely." *South Atlantic Quarterly* **106**(3): 445.

Cohen, P., Gilfoyle, T. and Horowitz, H. L. (2008). *The Flash Press: sporting male weeklies in 1840s New York*. Chicago, University of Chicago Press.

Collins, M. (1999). "The pornography of permissiveness: men's sexuality and women's emancipation in mid twentieth-century Britain." *History Workshop Journal* **47** (Spring): 99–120.

Colomina, B. (1994). *Privacy and Publicity*. Cambridge, MA, MIT Press.

Colomina, B. and Bloomer, J. (1992). *Sexuality & Space*. New York, Princeton Architectural Press.

Colosi, R. (2010). "A return to the Chicago school? From the 'subculture' of taxi dancers to the contemporary lap dancer." *Journal of Youth Studies* **13**(1): 1–16.

Colun, R. (2009). *Getting Life in Two Worlds: power and prevention in the New York City House Ball community*. New Brunswick, Rutgers University Graduate School.

Comella, L. (2008). "It's sexy. It's big business. And it's not just for men." *Contexts* **7**(3): 61–3.

—— (2010). "Repackaging sex: class, crass and the Good Vibrations model of sexual retail." In K. Hardy, S. Kingston and T. Sanders (eds) *New Sociologies of Sex Work*. London, Ashgate.

Conder, N. (1949). *Modern Architecture: an introduction*. London, Pellegrini & Cudahy.

Conover, P. (1975). "An analysis of communes and intentional communities with particular attention to sexual and genderal relations." *Family Coordinator* **24**(4): 453–64.

Constable, N. (2009). "The commodification of intimacy: marriage, sex, and reproductive labor." *Annual Review of Anthropology* **38**: 49–64.

Conway, D. (2004). "Every coward's choice? Political objection to military service in apartheid South Africa as sexual citizenship." *Citizenship Studies* **8**: 25–45.

Cook, M. (2003). *London and the Culture of Homosexuality, 1885–1914*. Cambridge, Cambridge University Press.

Corbin, A. (1990). *Women for Hire: prostitution and sexuality in France after 1850*. Cambridge, MA, Harvard University Press.

Couch, D. and Liamputtong, P. (2008). "Online dating and mating: the use of the internet to meet sexual partners." *Qualitative Health Research* **18**(2): 268.

Coulmont, B. (2007). *Sex-Shops: une histoire française*. Paris, Dilecta Press.

Coulmont, B. and Hubbard, P. (2010). "Consuming sex: socio-legal shifts in the space and place of sex shops." *Journal of Law and Society* **37**(1): 189–209.

Cowan, S. (2007). "Freedom and capacity to make a choice: a feminist analysis of consent in the criminal law of rape." In C. Stychin and V. Munro (eds) *Sexuality and the Law: feminist engagements*. London: Routledge-Cavendish Publishing.

Cox, R. (2007). "The au pair body: sex object, sister or student?" *European Journal of Women's Studies* **14**(3): 281–96.

Coy, M. (2009). "Milkshakes, lady lumps and growing up to want boobies: how the sexualisation of popular culture limits girls' horizons." *Child Abuse Review* **18**(6): 372–83.

Cressey, P. (1932). *The Taxi-dance Hall: a sociological study in commercialised recreation and city life*. Chicago, University of Chicago Press.

Cresswell, T. (1994). "Putting women in their place: the carnival at Greenham Common." *Antipode* **26**(1): 35–58.

—— (1998). "Night discourse: producing/consuming meaning on the street." In N. Fyfe (ed.) *Images of the Street*. London, Routledge.

—— (2000). "Mobility, syphilis, and democracy: pathologizing the mobile body." *Clio Medica/The Wellcome Series in the History of Medicine* **56**(1): 261–77.

Crofts, P. (2007). "Brothels and disorderly acts." *Public Space: the journal of law and social justice* **1**(2): 1–39.

Cross, S. (2005). "Paedophiles in the community: inter-agency conflict, news leaks and the local press." *Crime Media Culture* **1**(3): 284–300.

Cybriwsky, R. W. (2011). *Roppongi Crossing: the demise of a Tokyo nightclub district and the reshaping of a global city*. Atlanta, University of Georgia Press.

Daley, C. (2005). "From bush to beach: nudism in Australasia." *Journal of Historical Geography* **31**(1): 149–67.

Danta, D. (2009). "Ambiguous landscapes of San Pornando Valley." *Association of Pacific Coast Geographers* **71**: 15–30.

Davis, K. (2007). "The Bondi boys – un/Australian?" *Continuum* **21**(4): 501–10.

Day, K., Gough, B. and McFadden, M. (2004). "Warning! Alcohol can seriously damage your feminine health: a discourse analysis of recent British newspaper coverage of women and drinking." *Feminist Media Studies* **4**(2): 165–83.

Day, S. (2009). "Renewing the war on prostitution: the spectres of 'trafficking' and 'slavery'." *Anthropology Today* **25**(3): 1–3.

Dear, M. (1992). "Understanding and overcoming the NIMBY syndrome." *Journal of the American Planning Association* **58**(3): 288–300.

Del Casino Jr, V. (2007). "Flaccid theory and the geographies of sexual health in the age of Viagra." *Health & Place* **13**(4): 904.

Delany, S. (2001). *Times Square Red, Times Square Blue*. New York, New York University Press.

Deleuze, G. and Guattari, F. (1987). *A Thousand Plateaus: capitalism and schizophrenia*. Minneapolis, University of Minnesota Press.

Delph, E. (1978). *The Silent Community: public homosexual encounters*. Beverly Hills: Sage.

Dematteis, G. (2000). "Spatial images of European urbanisation." In A. Bagnasco and P. Le Gales (eds) *Cities in Contemporary Europe*. Cambridge, Cambridge University Press.

D'Emilio, J. (1998). *Sexual Politics, Sexual Communities: the making of a homosexual minority in the United States, 1940–1970*. Chicago, University of Chicago Press.

Denfeld, D. and Gordon, F. (1970). "The sociology of mate swapping: or the family that swings together clings together." *Journal of Sex Research* **6**(2): 85–100.

Dines, G. (2010). *Pornland: how porn hijacked our sexuality*. Boston, Beacon Press.

Doan, P. (2004). "Queers in the American city: transgendered perceptions of urban space." *Gender, Place and Culture* **14**(1): 57–74.

—— (2010). "The tyranny of gendered spaces: reflections from beyond the gender dichotomy." *Gender, Place and Culture* **17**(5): 635–54.

Doan, P. L. and Higgins, H. (2011). "The demise of queer space? Resurgent gentrification and the assimilation of LGBT neighborhoods." *Journal of Planning Education and Research*. **31**(1): online early DOI: 10.1177/0739456X10391266.

Doderer, Y. P. (2011). "LGBTQs in the city, queering urban space." *International Journal of Urban and Regional Research* **35**(2): 431–6.

Donzelot, J. (1997). *The Policing of Families*. New York, Random House.

D'Orlando, F. (2011). "The demand for pornography." *The Journal of Happiness Studies* **12**(1): 51–75.

Douglas, B. and Tewksbury, R. (2008). "Theaters and sex: an examination of anonymous sexual encounters in an erotic oasis." *Deviant Behavior* **29**(1): 1–17.

Douglas, M. (2002). *Purity and Danger: an analysis of concept of pollution and taboo*. London, Routledge.

Dreher, N. (1997). "The virtuous and the verminous: turn-of-the-century moral panics in London's public parks." *Albion: a quarterly journal concerned with British studies* **29**(2): 246–67.

Duggan, L. (2002). "The new homonormativity: the sexual politics of neoliberalism." In R. Castranova and D. Nelson (eds) *Materializing Democracy: toward a revitalized cultural politics*. Durham, Duke University Press.

Duncan, N. (1996). "Renegotiating gender and sexuality in public and private spaces." In N. Duncan (ed.) *Body Space*. London, Routledge.

Duncombe, J. and Marsden, N. (1993). "Love and intimacy: the gender division of emotion and emotion work: a neglected aspect of sociological discussion of heterosexual relationships." *Sociology* **27**(2): 221.

Edelman, B. (2009). "Red light states: who buys online adult entertainment?" *Journal of Economic Perspectives* **23**(1): 209–20.

Edelman, L. (1994). "Tearooms and sympathy; or, the epistemology of the water closet." In L. Edelman (ed.) *Homographies: essays in gay literary and cultural theory*. New York, Routledge.

—— (2004). *No Future: queer theory and the death drive*. Durham, Duke University Press.

Edwards, M. L. (2009). "Gender, social disorganization theory, and the locations of sexually oriented businesses." *Deviant Behavior* **31**(2): 135–58.

Elder, G. (2005). "Love for sale: marketing gay male p/leisure space in contemporary South Africa." In L. Nelson and J. Seager (eds) *A Feminist Companion to Geography*. Oxford, Blackwells.

Eldridge, A. and Roberts, M. (2008). "Hen parties: bonding or brawling?" *Drugs: education, prevention and policy* **15**(3): 323–8.

Elias, N. (1978). *The Civilizing Process: sociogenetic and psychogenetic investigations*. Oxford, Wiley-Blackwell.

Elwood, S. A. (2000). "Lesbian living spaces – multiple meanings of home." *Journal of Lesbian Studies* **4**(1): 11–27.

Emling, S. (2007). "Amsterdam to curb red light district." *Houston Chronicle*, 14 October: www.chron.com/disp/story.mpl/world/5210991.html (accessed 3 August 2011).

Engels, F. (1844). *The Condition of the Working Class in Britain*. Leipzig, Verlag Reclam.

England, K. (1993). "Suburban pink collar ghettos: the spatial entrapment of women?" *Annals of the Association of American Geographers* **83**(2): 225–42.

Engstrom, E. (2007). "Selling with sex in Sin City: the case of the Hard Rock Hotel casino." *Journal of Promotion Management* **13**(1): 169–88.

Evans, A., Riley, S. and Shankar, A. (2010). "Postfeminist heterotopias." *European Journal of Women's Studies* **17**(3): 211.

Evans, J. V. (2003). "Bahnhof boys: policing male prostitution in post-Nazi Berlin." *Journal of the History of Sexuality*, **12**(4): 605–34.

Eves, A. (2004). "Queer theory, butch/femme identities and lesbian space." *Sexualities* **7**(4): 480–96.

Farrar, J. (2002). *Opening Up: youth sex culture and market reform in Shanghai*. Chicago, Chicago University Press.

Felski, R. (1995). *The Gender of Modernity*. Cambridge, MA, Harvard University Press.

Ferreday, D. (2008). "'Showing the girl': the new burlesque." *Feminist Theory* **9**(1): 47.

Ferrell, R. (2000). "Copula: the logic of the sexual relation." *Hypatia* **15**(2): 100–14.

Florida, R. (2002). "The economic geography of talent." *Annals, Association of American Geographers* **92**(4): 743–55.

Florida, R. and Gates, G. (2005). "Tolerance and talent." In R. Florida (ed.) *Cities and the Creative Class*. New York, Routledge.

Forel, A. and Fetscher, R. (1931). *The Sexual Question*. Batavia, Geode Templerian.

Forest, B. (1995). "West Hollywood as symbol – the significance of place in gay identity." *Environment and Planning D: society and space* **13**(2): 133–57.

Foucault, M. (1977). *Discipline and Punish*. Harmondsworth, Penguin.

—— (1978). *The History of Sexuality Vol. 1: the will to knowledge*. London, Penguin.

—— (1982). *The History of Sexuality Vol. 2: the uses of pleasure*. London, Penguin.

—— (1990). *The History of Sexuality Vol. 3: care of the self*. London, Penguin.

Foucault, M., Bertani, M., Fontana, A. and Macey, D. (2003). *Society Must be Defended*. London, Picador.

Fougere, M. and Solitander, N. (2010). "Governmentality and the creative class: harnessing Bohemia, diversity and freedom for competitiveness." *International Journal of Management Concepts and Philosophy* **4**(1): 41–59.

Fox, K., Nobles, M. R. and Piquero, A. R. (2009). "Gender, crime victimization and fear of crime." *Security Journal* **22**(1): 24–39.

Frances, R. (1994). "The history of female prostitution in Australia." In R. Perkins, G. Prestage, R. Sharp and F. Lovejoy (eds) *Sex Work and Sex Workers in Australia*. Sydney, University of New South Wales Press.

Franck, K. and Stevens, Q. (2007). *Loose Space: possibility and diversity in urban life*. London, Routledge.

Frank, K. (2008). "Not gay, but not homophobic: male sexuality and homophobia in the lifestyle." *Sexualities* **11**(4): 435.

Frankis, J. (2009). "Public sexual cultures: a systematic review of qualitative research investigating men's sexual behaviors with men in public spaces." *Journal of Homosexuality* **56**(7): 861–93.

Fraterrigo, E. (2008). "The answer to suburbia: Playboy's urban lifestyle." *Journal of Urban History* **34**(5): 747–74.

Friedman, A. (2000). *Prurient Interests: gender, democracy, and obscenity in New York City, 1909–1935*. New York, Columbia University Press.

Frisby, D. (2001). *Cityscapes of Modernity: critical explorations*. Cambridge, Polity.

Fuss, D. (1989). *Essentially Speaking: feminism, nature and difference*. London, Routledge and Kegan Paul.

Gaffney, J. and Beverley, K. (2001). "Contextualizing the construction and social organization of the commercial male sex industry in London at the beginning of the twenty-first century." *Feminist Review* **67**: 133–41.

Gaissad, L. (2005). "From nightlife conventions to daytime hidden agendas: dynamics of urban sexual territories in the south of France." *Journal of Sex Research* **42**(1): 20–7.

Giddens, A. (1992). *The Transformation of Intimacy: sexuality, love and eroticism in modern societies*. New York, Stanford University Press.

Giffney, N. (2004). "Studies denormatizing queer theory: more than (simply) lesbian and gay." *Feminist Theory* **5**(1): 73–8.

Godden, L. (2001). "The bounding of vice: prostitution and planning law." *Griffith Law Review* **10**(1): 77–98.

Goodey, J. (2003). "Migration, crime and victimhood: responses to trafficking in the EU." *Punishment and Society* **5**(4): 415–31.

Gorman-Murray, A. (2006). "Gay and lesbian couples at home: identity work in domestic space." *Home Cultures* **3**: 145–67.

—— (2007). "Contesting domestic ideals: queering the Australian home." *Australian Geographer* **38**(2): 195–213.

Gorman-Murray, A. and Waitt, G. (2009). "Queer-friendly neighbourhoods: interrogating social cohesion across sexual difference in two Australia neighbourhoods." *Environment and Planning A* 41: 2855–73.

Gorman-Murray, A., Waitt, G. and Gibson, C. (2008). "A queer country? A case study of the politics of gay/lesbian belonging in an Australian country town." *Australian Geographer* **39**(2): 171–91.

Grazian, D. (2008). *On the Make: the hustle of urban nightlife*. Chicago, Chicago University Press.

Green, A., Follert, M., Osterland, K. and Paquin, J. (2008). "Space, place and sexual sociality: towards an 'atmospheric analysis'." *Gender, Work, Organization* **17**: 7–27.

Green, E. and Singleton, C. (2006). "Risky bodies at leisure: young women negotiating space and place." *Sociology* **40**(5): 853–71.

Grosz, E. (1994). *Volatile Bodies: toward a corporeal feminism*. Gary, Indiana University Press.

Groves, J., Newton, D. C., Chen, M. Y., Hocking, J., Bradshaw, C. S., Fairley, C. K. (2008). "Sex workers working within a legalised industry: their side of the story." *Sexually Transmitted Infections* **84**(5): 393–4.

Grubesic, T., Murray, A. T. and Mack, E. A. (2008). "Sex offenders, housing and spatial restriction zones." *GeoJournal* **73**(4): 255–69.

Gulf News (2009). "Caught in th act: Dubai beach indecency." Online. Available http://gulfnews.com/news/gulf/uae/crime/caught-in-the-act-dubai-beach-indecency-1.55448 (accessed 5 September 2011).

Gunn, S. (2002). "City of mirrors: the Arcades Project and urban history." *Journal of Victorian Culture* **7**(2): 263–75.

Guyatt, V. (2005). "Gender performances in a service orientated workplace in Aotearoa/New Zealand." *New Zealand Geographer* **61**(3): 203–12.

Hadfield, P. and Measham, F. (2009). "England and Wales." In P. Hadfield (ed.) *Nightlife and Crime: social order and governance in international perspective*. Oxford, Oxford University Press.

Halberstam, J. (2005). *In a Queer Time and Place: transgender bodies, subcultural lives*. New York, NYU Press.

Hall, L. (1992). "Forbidden by God, despised by men: masturbation, medical warnings, moral panic, and manhood in Great Britain, 1850–1950." *Journal of the History of Sexuality* **2**(3): 365–87.

Hall, T. (2007). "Rent-boys, barflies, and kept men: men involved in sex with men for compensation in Prague." *Sexualities* **10**(4): 457.

Halperin, D. (1986). "One hundred years of homosexuality." *Diacritics* **16**(2): 34–45.

—— (2002). "Sex before sexuality: pedastry, politics and power in classical Athens." In L. Alcoff (ed.) *Identities: race, class, gender and nation*. Oxford, Blackwells.

—— (2003). "The normalizing of queer theory." *Journal of Homosexuality* **45**(2): 339–43.

Hammers, C. (2009). "An examination of lesbian/queer bathhouse culture and the social organization of (im)personal sex." *Journal of Contemporary Ethnography* **38**(3): 308.

Handyside, F. (2007). "It's either fake or foreign: the cityscape in Sex and the City." *Continuum* **21**(3): 405–18.

Hanna, J. L. (2005). "Exotic dance adult entertainment: a guide for planners and policy makers." *Journal of Planning Literature* **20**: 116–34.

Harcourt, C., Egger, S. and Donovan, B. (2005). "Sex work and the law." *Sexual Health* **2**(1): 121–28.

Hareven, T. (1976). "Modernization and family history: perspectives on social change." *Signs*: 190–206.

Harris, D. (1997). *The Rise and Fall of Gay Culture*. New York, Hyperion Press.

Harsin, J. (1985). *Policing Prostitution in Nineteenth-Century Paris*. New York, Princeton University Press.

Hart, K. (2004). "We're here, we're queer and we're better than you: the representational superiority of gay men to heterosexuals on Queer Eye for the Straight Guy." *The Journal of Men's Studies* **12**(3): 241–53.

Haskey, J. (2005) "Living arrangements in contemporary Britain: having a partner who usually resides elsewhere and living apart together." *Population Trends* **122**: 35–45.

Hastings, S. and Magowan, F. (2010). *Transgressive Sex: subversion and control in erotic encounters*. New York: Berghahn Books.

Hausbeck, K. and Brents, B. (2010). "Nevada's legal brothels." In R.Weitzer (ed.) *Sex for Sale: prostitution, pornography and the sex industry*. New York, Routledge.

Hawkes, G. and Egan, R. (2008). "Landscapes of erotophobia: the sexual(ized) child in the postmodern anglophone West." *Sexuality & Culture* **12**(4): 193–203.

Hayden, D. (1980). "What would a non-sexist city be like? Speculations on housing, urban design, and human work." *Signs* **5**(3): 170–87.

He, T. (2007). "Cyberqueers in Taiwan: locating histories of the margins." *International Journal of Women's Studies* **8**(2): 55–73.

Heap, C. (2003). "The city as a sexual laboratory: the queer heritage of the Chicago School." *Qualitative Sociology* **26**(4): 457–87.

Hearns, J. (2006). "The implications of information and communication technologies for sexualities and sexualised violences: contradictions of sexual citizenships." *Political Geography* **25**(8): 944–63.

Hemingway, J. (2006). "Sexual learning and the seaside: relocating the 'dirty weekend' and teenage girls' sexuality." *Sex Education: Sexuality, Society and Learning* **6**(4): 429–43.

Hemmings, C. (2002). *Bisexual spaces: a geography of sexuality and gender*. London, Routledge.

Hennelly, S. (2010). "Public space, public morality: the media construction of sex in public places." *Liverpool Law Review* **31**(1): 69–91.

Hennig, J. (2006). *Trafficking in Human Beings and the 2006 World Cup in Germany*. Geneva, International Organization for Migration.

Herald, M. (2004). "A bedroom of one's own: morality and sexual privacy after Lawrence v. Texas." *Yale Journal of Law and Feminism* **16**: 1.

Herdt, G. (ed.) (2009). *Moral Panics, Sex Panics*. New York, New York University Press.

Heynen, H. (2005). "Modernity and domesticity." In H. Heynen and G. Baydar (eds) *Negotiating Domesticity: spatial productions of gender in modern architecture*. London, Routledge.

Hickey, A. (2002). "Between two spheres: comparing state and federal approaches to the right to privacy and prohibitions against sodomy." *Yale Law Journal* **111**(4): 993–1030.

Higgs, D. (ed.) (1999). *Queer Sites: gay urban histories since 1600*. London, Routledge.

Himmelfarb, G. (1995). *The De-Moralization of Society: from Victorian virtues to modern values*. Berlin, Alfred Knopf.

Hindle, P. (1994). "Gay communities and gay spaces in the city." In S. Whittle (ed.) *On the Margins of the City*. Manchester, Arena Press.

Hobbs, D., O'Brien, K. and Westmarland, L. (2007). "Connecting the gendered door: women, violence and doorwork." *British Journal of Sociology* **58**(1): 21–38.

Hofmann, S. (2010). "Corporeal entrepreneurialism and neoliberal agency in the sex trade at the US-Mexican border." *WSQ: Women's Studies Quarterly* **38**(3–4): 233–56.

Hoigard, C. and Finstad, L. (1992). *Backstreets: prostitution, money, and love*. Philadelphia, Pennsylvania State University Press.

Holgersson, C. and Svanstrom, Y. (2004). *Lagiga och Oligaga Affarer – om Sexkopoch Organizartioner*. Stockholm, NorFa.

Holland, E. (1999). *Deleuze and Guattari's Anti-Oedipus: introduction to schizo-analysis*. New York, Routledge.

Holmes, D., O'Byrne, P. and Gastaldo, G. (2007). "Setting the space for sex: architecture, desire and health issues in gay bathhouses." *Journal of Nursing Studies* **44**: 273–84.

Holmes, J. (2006). “Bare bodies, beaches, and boundaries: abjected outsiders and rearticulation at the nude beach.” *Sexuality & Culture* **10**(4): 29–53.

Home Office (2006). *A Coordinated Prostitution Strategy and a Summary of Responses to Paying the Price*. London, Home Office.

Hooper, B. (1998). “The poem of male desires.” In L. Sandercock (ed.) *Making the Invisible Visible: a multicultural planning history*. Los Angeles, University of California Press.

Hornsey, R. (2010). *The Spiv and the Architect: unruly life in post war London*. Minneapolis, University of Minnesota Press.

Houlbrook, M. (2000). “The private world of public urinals: London 1918–57.” *London Journal* **25**: 52–70.

—— (2006). “Cities”. In M. Houlbrook and. H. Cocks (eds) *The Modern History of Sexuality*. London, Macmillan.

Howard, J. (1995). “The library, the park, and the pervert: public space and homosexual encounter in post-World War II Atlanta”. *Radical History Review* **62**: 166–87.

Howell, P. (2001). “Sex and the city of bachelors: sporting guidebooks and urban knowledge in nineteenth-century Britain and America.” *Cultural Geographies* **8**(1): 20.

—— (2004). “Race, space and the regulation of prostitution in colonial Hong Kong.” *Urban History* **31**(2): 229–46.

—— (2009). *Geographies of Regulation: policing prostitution in nineteenth century Britain and the Empire*. Cambridge, Cambridge University Press.

Howell, P., Beckingham, D. and Moore, F. (2008). “Managed zones for sex workers in Liverpool: contemporary proposals, Victorian parallels.” *Transactions of the Institute of British Geographers* **33**(2): 233–50.

Hubbard, P. (1997). “Red light districts and toleration zones: geographies of female street prostitution in England and Wales.” *Area* **29**(2): 129–40.

—— (1998). “Community action and the displacement of street prostitution: evidence from British cities.” *Geoforum* **29**(3): 269–86.

—— (1999). *Sex and the City*. Chichester, Ashgate.

—— (2000). “Desire/disgust: mapping the moral contours of heterosexuality.” *Progress in Human Geography* **24**(2): 191–217.

—— (2003). “A good night out? Multiplex cinemas as sites of embodied leisure.” *Leisure Studies* **22**(3) 255–72.

—— (2004). “Revenge and injustice in the neoliberal city: uncovering masculinist agendas.” *Antipode* **36**(4): 665–86.

—— (2005). “Accommodating Otherness: anti-asylum centre protest and the maintenance of white privilege.” *Transactions of the Institute of British Geographers* **30**(1): 52–65.

—— (2006). *Key ideas in Human Geography: the city*. London, Routledge.

—— (2008). “Here, there, everywhere: the ubiquitous geographies of heteronormativity.” *Geography Compass*. **2**: 640–58.

—— (2009). “Opposing striptopia: the embattled spaces of adult entertainment.” *Sexualities* **12**(6): 721–45.

Hubbard, P. and Lilley, K. (2004). "Pacemaking the modern city: the urban politics of speed and slowness." *Environment and Planning D: society and space* **22**(2): 273–94.

Hubbard, P. and Sanders, T. (2003). "Making space for sex work: female street prostitution and the production of urban space." *International Journal of Urban and Regional Research* **27**(1): 75–87.

Hubbard, P. and Whowell, M. (2008). "Revisiting the red light district: still neglected, immoral and marginal?" *Geoforum* **39**(5): 1743–55.

Hubbard, P., Campbell, R., O'Neill, M. and Scoular, J. (2007). "Prostitution, gentrification, and the limits of neighbourhood space." In R. Atkinson and G. Helms (eds) *Securing an Urban Renaissance: crime, community, and British urban policy*. Bristol, Policy Press.

Hubbard, P., Matthews, R. and Scoular, J. (2008). "Regulating sex work in the EU: prostitute women and the new spaces of exclusion." *Gender, Place and Culture* **15**(2): 137–52.

—— (2009). "Controlling sexually oriented businesses: law, licensing and the geographies of a controversial land use." *Urban Geography* **30**(2): 185–205.

Hubbard, P., Matthews, R., Scoular, J. and Agustin, L. (2008). "Away from prying eyes? The urban geographies of 'adult entertainment'." *Progress in Human Geography* **32**(3): 363–81.

Hughes, D. (2002). "The use of new communication and information technologies for the sexual exploitation of women and children." *Hastings Law Review* **13**(1): 127–46.

Hughes, H. L. (2003). "Marketing gay tourism in Manchester: new market for urban tourism or destruction of 'gay space'?" *Journal of Vacation Marketing* **9**(2): 152–63.

Humphreys, L. (1970). "Tearoom trade." *Society* **7**(3): 10–25.

Hunt, A. (2002). "Regulating heterosocial space: sexual politics in the early twentieth century." *Journal of Historical Sociology* **15**(1): 1–34.

Hunt, L. (1996). *The Invention of Pornography: obscenity and the origins of modernity, 1500–1800*. New York, Zone Books.

Hunter, I., Saunders, D. and Williamson, D. (1993). *On Pornography: literature, sexuality and obscenity law*. London, Macmillan and New York, St Martin's Press.

Hutton, F. (2004). "Up for it, mad for it? Women, drug use and participation in club scenes." *Health, Risk & Society* **6**(3): 223–37.

Illouz, E. (1997). *Consuming the Romantic Utopia: love and the cultural contradictions of capitalism*. Los Angeles, University of California Press.

Ingram, G. B. (2007). *Globalizing Homosexual and Male Guest Worker Identities: the strategic role of Dubai's open beach*. Unpublished paper from Sexuality and Space conference, San Francisco.

Insight Associates (1994). *Report on the Secondary Effects of the Concentration of Adult Use Establishments in the Times Square Area*. New York, Times Square Business Improvement District.

Irvine, J. (2007). "Transient feelings: sex panics and the politics of emotions." *GLQ: a journal of lesbian and gay studies* **14**(1): 1–41.

Isherwood, C. (1939). *Goodbye to Berlin*. London, Hogarth Press.

Iveson, K. (2007). *Publics and the City*. Cambridge, Wiley-Blackwell.

Jackson, C. and Tinkler, P. (2007). "'Ladettes' and 'modern girls': 'troublesome' young femininities." *Sociological Review* **55**(2): 251–72.

Jackson, P. A. (1997). 'Kathoey – gay – man: the historical emergence of gay male identity in Thailand'. In L. Jolly and M. Manderson (eds) *Sites of Desire/ Economies of Pleasure, Sexualities in Asia and the Pacific*. Chicago, University of Chicago Press.

—— (2009). "Capitalism and global queering: national markets, parallels among sexual cultures and multiple queer modernities." *GLQ: a journal of lesbian and gay studies* **15**(3): 357–95).

Jackson, S. (2009). *Materialist Feminism, the Pragmatist Self and Global Late Modernity*. Proceedings from GEXcel Theme 1: Gender, Sexuality and Global Change, Linkoping University.

Jackson, S. and Scott, S. (2004). "Sexual antinomies in late modernity." *Sexualities* **7**(2): 233.

Jacobs, J. (2009). "Have sex will travel: romantic 'sex tourism' and women negotiating modernity in the Sinai." *Gender, Place and Culture* **16**(1): 43–61.

Jacobs, K. (2004). "Pornography in small places and other spaces." *Cultural Studies* **18**(1): 67–83.

—— (2008). "Porn arousal and gender morphing in the twilight zone." In K. Jacobs (ed.) *C'lick Me: a netporn studies reader*: www.networkcultures.org/_uploads/24.pdf (accessed 6 May 2011).

Jagose, A. (1997). *Queer Theory: an introduction*. New York, New York University Press.

—— (2009). "Feminism's queer theory." *Feminism & Psychology* **19**(2): 157.

Jamieson, A. (2004). "Capital crackdown on stag and hen party binge drinkers is on the cards." Online. Available http://news.scotsman.com/edinburgh/capital-crackdown-on-stag-and.2514634.jp (accessed 3 August 2011).

Jancovich, M., and Faire, L. (2003). *The Place of the Audience: cultural geographies of film consumption*. London, British Film Institute.

Jarvis, J. and Kallas, P. (2008). "Estonian tourism and the accession effect: the impact of European Union membership on the contemporary development patterns of the Estonian tourism industry." *Tourism Geographies* **10**(4): 474–94.

Jayne, M., Holloway, S. and Valentine, G. (2006). "Drunk and disorderly: alcohol, urban life and public space." *Progress in Human Geography* **20**(4): 451–68.

Jeffreys, S. (2008). *The Industrial Vagina: the political economy of the global sex trade*. London, Routledge.

Jenks, R. (1998). "Swinging: a review of the literature." *Archives of Sexual Behavior* **27**(5): 507–21.

Jennings, R. (2007). *A Lesbian History of Britain: love and sex between women since 1500*. London, Greenwood World Publishing.

Jensen, R. and Dines, G. (1998). "The content of mass-marketed pornography." In G. Dines, R. Jensen and A. Russo (eds) *Pornography: the production and consumption of inequality*. London, Routledge.

Jervis, J. (1999). *Transgressing the Modern: explorations in the western experience of otherness*. Oxford, Wiley-Blackwell.

Jewkes, Y. (2010). "Much ado about nothing? Representations and realities of online soliciting of children." *Journal of Sexual Aggression: an international, interdisciplinary forum for research, theory and practice* **16**(1): 5–18.

Jeyasingham, D. (2010). "Building heteronormativity: the social and material reconstruction of men's public toilets as spaces of heterosexuality." *Social & Cultural Geography* **11**(4): 307–25.

Johnson, J. A. (2010). "To catch a curious clicker?" In Boyle, K. (ed.) *Everyday Pornography*. London, Routledge.

Johnson, P. (2007). "Ordinary folk and cottaging: law, morality, and public sex." *Journal of Law and Society* **34**(4): 520–43.

—— (2010). "Law, morality and disgust: the regulation of 'extreme pornography' in England and Wales." *Social and Legal Studies* **19**(2): 147–64.

Johnston, L. (2001). "(Other) bodies and tourism studies." *Annals of Tourism Research* **28**(1): 180–201.

—— (2005). *Queering Tourism: paradoxical performances at gay pride parades*. London, Routledge.

—— (2007). "Mobilizing pride/shame: lesbians, tourism and parades." *Social & Cultural Geography* **8**: 29–45.

—— (2010). "Sites of excess: the spatial politics of touch for drag queens in Aotearoa New Zealand." *Emotion, Space and Society*. Forthcoming.

Johnston, L. and Longhurst, R. (2008). "Queer(ing) geographies 'down under': some notes on sexuality and space in Australasia." *Australian Geographer* **39**(3): 247–57.

—— (2010). *Space, Place and Sex: geographies of sexualities*. Lanham, Rowman and Littlefield.

Johnston, L. and Valentine, G. (1995). "Wherever I lay my girlfriend, that's my home: the performance and surveillance of lesbian identities in domestic environments." In D. Bell and G. Valentine (eds) *Mapping Desire: geographies of sexualities*. London, Routledge.

Juffer, J. (1998). *At Home with Pornography, Women, Sex, and Everyday Life*. New York, New York University Press.

Kantola, J. and Squires, D. (2004). "Discourses surrounding prostitution policies in the UK." *European Journal of Women's Studies* **11**(1): 77.

Kates, S. M. and Belk, R. W. (2001). "The meanings of lesbian and gay pride day: resistance through consumption and resistance to consumption." *Journal of Contemporary Ethnography* **30**(4): 392–429.

Kavanaugh, P. and Anderson, T. (2009). "Managing physical and sexual assault risk in urban nightlife: individual- and environmental-level influences." *Deviant Behavior* **30**(8): 680–714.

Kavaratzis, M. and Ashworth, G. (2007). "Coffeeshops, canals and commerce: marketing the city of Amsterdam." *Cities* **24**: 16–25.

Keire, M. L. (2010). *For Business & Pleasure: red-light districts and the regulation of vice in the United States, 1890–1933*. Baltimore, John Hopkins University Press.

Kelly, E. D. and Cooper, C. (2000). *Everything You Always Wanted to Know About Regulating Sex Businesses*. Chicago: APA.

Kelly, P. (2008). *Lydia's Open Door: inside Mexico's most modern brothel*. Los Angeles, University of California Press.

Kendrick, W. (1996). *The Secret Museum: pornography in modern culture*. Los Angeles, University of California Press.

Kent, T. and Brown, R. B. (2006). "Erotic retailing in the UK (1963–2003)." *Journal of Management History* **12**(2): 199–211.

Kentlyn, S. (2008). "The radically subversive space of the queer home: safety house and neighbourhood watch." *Australian Geographer* **39**(3): 327–37.

Kerkin, K. (2004). "Discourse, representation and urban planning: how a critical approach to discourse helps reveal the spatial re-ordering of street sex work." *Australian Geographer* **35**(2): 185–92.

Kern, L. (2007). "Reshaping the boundaries of public and private life: gender, condominium development, and the neoliberalization of urban living." *Urban Geography* **28**(7): 657–81.

Kinnell, H. (2008). *Violence and Sex Work in Britain*. Cullompton, Willan.

Kipnis, L. (1996). *Bound and Gagged: pornography and the politics of fantasy*. Durham, Duke University Press.

Kirby, S. and Hay, I. (1997). "(Hetero)sexing space: gay men and 'straight' space in Adelaide, South Australia." *Professional Geographer* **49**(3): 295–305.

Kirkey, K. and Forsyth, A. (2001). "Men in the valley: gay male life on the suburban-rural fringe." *Journal of Rural Studies* **17**(4): 421–41.

Kitchin, R. (2002). "Sexing the city: the sexual production of non-heterosexual space in Belfast, Manchester and San Francisco." *City* **6**(2): 205–18.

Kitzinger, C. and Wilkinson, S. (1994). "Virgins and queers: rehabilitating heterosexuality?" *Gender and Society* **8**(3): 444–62.

Klein, F. and Anderlini-D'Onofrio, S. (2005). *Plural Loves: designs for bi and poly living*. London, Routledge.

Klesse, C. (2006). "Polyamory and its 'Others': contesting the terms of non-monogamy." *Sexualities* **9**(5): 565.

Kneale, J. (1999). "A problem of supervision: moral geographies of the nineteenth-century British public house." *Journal of Historical Geography* **25**(3): 333–48.

Kneeland, G. and Davis, J. (1917). *Commercialized Prostitution in New York City*. New York, The Century Company.

Knopp, L. (1992). "Sexuality and the spatial dynamics of capitalism." *Environment and Planning D: society and space* **10**: 651–69.

—— (1995). "If you're going to get all hyped up you'd better go somewhere." *Gender, Place and Culture* **2**(1): 85–8.

—— (1997a). "Gentrification and gay neighborhood formation in New Orleans." In N. Glukman and B. Reed (eds) *Homo Economics: capitalism, community, and lesbian and gay life*. London, Routledge.

—— (1997b). "Rings, circles and perverted justice: gay judges and moral panic in contemporary Scotland." In M. Keith and S. Pile (eds) *Geographies of Resistance*. London, Routledge.

—— (1998). "Sexuality and urban space: gay male identities, communities and cultures in the U.S., U.K. and Australia." In R. Fincher and J. Jacobs (eds) *Cities of Difference*. New York, Guilford.

—— (2007). "From lesbian to gay to queer geographies: pasts, prospects and possibilities." In G. Brown, K. Browne and J. Lim (eds) *Geographies of Sexualities: theory practices and politics*. Chichester, Ashgate.

Kolvin, P. (2010). *Sex Licensing*. London, Institute of Licensing.

Kong, T. (2006). "What it feels like for a whore: the body politics of women performing erotic labour in Hong Kong." *Gender, Work & Organisation* **13**(5): 509–34.

—— (2009). "More than a sex machine: accomplishing masculinity among Chinese male sex workers in the Hong Kong sex industry." *Deviant Behavior* **30**(8): 715–45.

Koskela, H. (2004). "Webcams, TV shows and mobile phones: empowering exhibitionism." *Surveillance & Society* **2**(2/3): 199–215.

Kracauer, S. (1932). "Wiederholung." *Frankfurter Zeitung* 29 May.

Kramer, L. S. (1988). *Threshold of a New World: intellectuals and the exile experience in Paris, 1830–1848*. Ithaca, Cornell University Press.

Kristeva, J. (1982). *Powers of Horror: an essay on abjection*. New York, Columbia University Press.

Kuhanen, J. (2010). "Sexualised space, sexual networking and the emergence of AIDS in Rakai, Uganda." *Health & Place* **16**(2): 226–35.

Kunkel, J. (2011). "These dolls are an attraction." In J. Kunkel and M. Mayer (eds) *Urban Restructuring and Flexible Neoliberalization*. London, Palgrave Macmillan.

Kutz-Flamenbaum, R. (2007). "Code pink, raging grannies, and the missile dick chicks: feminist performance activism in the contemporary anti-war movement." *NWSA Journal* **19**(1): 89.

Laite, J. (2008). "The Association for Moral and Social Hygiene: abolitionism and prostitution law in Britain (1915–59)." *Women's History Review* **17**(2): 207–23.

—— (2009). "Historical perspectives on industrial development, mining and prostitution." *The Historical Journal* **52**: 739–61.

Langhamer, C. (2007). "Love and courtship in mid-twentieth-century England." *The Historical Journal* **50**(1): 173–96.

Langman, L. (2008). "Punk, porn and resistance: carnivalisation and the body in everyday life." *Critical Sociology* **56**(4): 657–77.

Larsen, E. (1992). "The politics of prostitution control: interest group politics in four Canadian cities." *International Journal of Urban and Regional Research* **16**(2): 169–89.

Lauria, M. and Knopp, L. (1985). "Towards an analysis of the role of gay communities in the urban renaissance" *Urban Geography* **6**(2): 152–69.

Leader, D. and Groves, J. (2005). *Introducing Lacan*. London, Icon.

Leap, W. (1999). *Public Sex/Gay Space*. New York, Columbia University Press.

Leary, D. and Minichiello, V. (2007). "Exploring the interpersonal relationships in street-based male sex work: results from an Australian qualitative study." *Journal of Homosexuality* **53**(1–2): 75–110.

Legg, S. (2009). "An intimate and imperial feminism: Meliscent Shephard and the regulation of prostitution in colonial India." *Environment and Planning D: society and space* **28**(1): 68–94.

Le Monnier, Y. and Cousin, A. (2004). *Guide du Paris Sexy 2004*. Paris, Musserdine.

Leslie, D. and Reimer, S. (2003). "Gender, modern design, and home consumption." *Environment and Planning D: society and space* **21**(3): 293–316.

Levine, M. (1979). "Gay ghetto." *Journal of Homosexuality* **4**: 363–77.

Levine, P. (2003). *Prostitution, Race, and Politics: policing venereal disease in the British Empire*. London, Routledge.

Lewis, C. and Pile, S. (1996). "Woman, body, space: Rio carnival and the politics of performance." *Gender, Place and Culture – A Journal of Feminist Geography* **3**: 23–42.

Liepe Levinson, K. (2002). *Strip Show: performances of gender and desire*. New York, Routledge.

Lim, K. F. (2004). "Where love dares (not) speak its name: the expression of homosexuality in Singapore." *Urban Studies* **41**(9): 1759–88.

Lin, H. S. (2009). "Private love in public space: love hotels and the transformation of intimacy in contemporary Japan." *Asian Studies Review* **32**(1): 31–56.

Linz, D., Paul, B., Land, K. C., Ezell, M. E. and Williams, J. R. (2004). "An examination of the assumption that adult businesses are associated with crime in surrounding areas: a secondary effects study in Charlotte, North Carolina." *Law and Society Review* **38**: 69–104.

Lisker, J. (1969). "Homo nest raided, queen bees are stinging mad." *New York News*, 6 July.

Listerborn, C. (2004). "Prostitution as urban radical chic: the silent acceptance of female exploitation." *City* **7**(2): 237–45.

Little, J. (2003). "'Riding the rural love train': heterosexuality and the rural community." *Sociologia Ruralis* **43**(4): 401–17.

Lloyd, J. (2008). "Home alone: selling new domestic spaces". In A. Cronin and K. Hetherington (eds) *Consuming the Entrepreneurial City: image, memory and spectacle*. London, Routledge.

Lloyd, J. and Johnson, J. (2004). "Dream stuff: the postwar home and the Australian housewife, 1940–60." *Environment and Planning D: society and space* **22**(2): 251–72.

Löw, M. and Ruhne, R. (2009). "Domesticating prostitution: study of an interactional web of space and gender." *Space and Culture* **12**: 232–49.

Lowman, J. (1992). "Street prostitution control: some Canadian reflections on the Finsbury Park experience." *British Journal of Criminology* **32**: 1–17.

—— (2009). "Violence and the outlaw status of (street) prostitution in Canada." In D. Canter, M. Ioannou and D. Youngs (eds) *Safer Sex in the City: the experience and management of street prostitution.* Chichester, Ashgate.

Luamann, E. O., Ellingson, S., Mahay, J., Paik, A. and Youm, Y. (2004). *The Sexual Organisation of the City.* Chicago, Chicago University Press.

Lumby, C., Allbury, K. and McKee, A. (2008). *The Porn Report.* Melbourne, Melbourne Unversity Press.

McCalman, I. (1988). *Radical Underworld: prophets, revolutionaries, and pornographers in London, 1795–1840.* Cambridge, Cambridge University Press.

McCleary, R. and Weinstein, A. C. (2009). "Do 'off-site' adult businesses have secondary effects? Legal doctrine, social theory, and empirical evidence." *Law & Policy* **31**(2): 217–35.

McDowell, L. (1983). "Towards an understanding of the gender division of urban space." *Environment and Planning D: society and space* **1**(1): 59–72.

McGhee, D. (2004). "Beyond toleration: privacy, citizenship and sexual minorities in England and Wales." *The British Journal of Sociology* **55**(3): 357–75.

McGhee, D. and Moran, L. (2000). "Perverting London: the cartographic practices of law." *Law and Critique* **9**(2): 207–24.

McGinley, A. C. (2007). "Harassing 'girls' at the Hard Rock: masculinities in sexualized environments." *University of Illinois Law Review* Paper 20: http://scholars.law.unlv.edu/facpub/20 (accessed 6 May 2011).

McGrath, M. (2010). *Girls! Girls! Girls! Next Door: neighbourhood responses to striptease venues in Portland, Oregon.* Association of Amercian Geographers Annual Conference. Washington, DC.

McKee, A., McKee, A., Albury, K., Dunne, M., Grieshaber, S., Hartley, J., Lumby, C. and Mathews, B. (2010). "Healthy sexual development: a multidisciplinary framework for research." *International Journal of Sexual Health* **22**(1): 14–19.

McKeganey, N. and Barnard, M. (1996). *Sex Work on the Streets.* Buckingham, Open University Press.

McKewon, E. (2003). "The historical geography of prostitution in Perth, Western Australia." *Australian Geographer* **34**(3): 297–310.

MacKinnon, C. (1993). *Only Words.* Cambridge, MA, Harvard University Press.

—— (1994). "Vindication and resistance: a response to the Carnegie Mellon study of pornography in cyberspace." *Georgia Law Journal* **83**: 1959.

McKinstry, W. (1974). "The pulp voyeur: a peek at pornography in public places." *Deviance: field studies and self-disclosures*: 30–40.

McNair, B. (2002). *Striptease Culture: sex, media and the democratization of desire.* London, Routledge.

McNeill, D. (2008). "The hotel and the city." *Progress in Human Geography* **32**(3): 383–98.

McRobbie, A. (2008). *The Aftermath of Feminism: gender, culture and social change.* London, Routledge.

McQuire, S. (2005). "Immaterial architectures: urban space and electric light." *Space and Culture* **8**(2): 126–35.

Maddison, S. (2002). "Small towns, boys and ivory towers: a naked academic." In J. Campbell and J. Harbord (eds) *Temporalities, Autobiography and Everyday Life*. Manchester, Manchester University Press.

—— (2010). "Online obscenity and myths of freedom." In F. Attwood (ed.) *Porn. com: making sense of online pornography*. New York, Peter Lang Publishing.

Mahon-Daly, P. and Andrews, A. G. (2002). "Liminality and breastfeeding: women negotiating space and two bodies." *Health & Place* **8**(2): 61–76.

Mahood, L. (1990). "The Magdalene's friend: prostitution and social control in Glasgow, 1869–90." *Women's Studies International Forum* **13**(1): 49–61.

Mai, N. (2009). *Migrant Workers in the UK Sex Industry*. London, Institute for the Study of European Transformation.

Mai, N. and King, R. (2009). "Love, sexuality and migration: mapping the issue(s)." *Mobilities* **4**(3): 295–307.

Malbon, B. (1999). *Clubbing: dancing, ecstasy and vitality*. London, Routledge.

Malecki, E. J. and Ewers, M. C. (2007). "Labor migration to world cities: with a research agenda for the Arab Gulf." *Progress in Human Geography* **31**(4): 467–84.

Mallett, S. (2004). "Understanding home: a critical review of the literature." *Sociological Review* **52**(1): 62–89.

Manchester, C. (1986). *Sex Shops and the Law*. London, Gower.

Manderson, L. and Jolly, M. (1997). *Sites of Desire, Economies of Pleasure: sexualities in Asia and the Pacific*. Chicago, University of Chicago Press.

Markus, T. (1993). *Buildings and Power: freedom and control in the origin of modern building types*. London, Routledge.

Markwell, K. (2002). "Mardi Gras Tourism and the construction of Sydney as an international gay and lesbian city." *GLQ: a journal of lesbian and gay studies* **8**(1–2): 81–99.

Marlet, G. and van Woerkens, C. (2005). *Tolerance, Aesthetics, Amenities or Jobs? Dutch city attraction to the creative class*. Discussion paper series 05–33. Utrecht, Tjalling C. Koopmans Research Institute.

Marriott, S. (2009). "The audience of one: adult chat television and the architecture of participation." *Screen* **50**(1) 25–34.

Marston, S. (2004). "A long way home: domesticating the social production of scale." In E. Sheppard and R. B. McMaster (eds) *Scale and Geographic Inquiry*. Oxford, Blackwell.

Marttila, A. M. (2008). "Desiring the 'Other': prostitution clients on a transnational red-light district in the border area of Finland, Estonia and Russia." *Gender Technology and Development* **12**(1): 31–51.

Mason, J. (2005). "'Affront or alarm': performance, the law and the 'female breast' from Janet Jackson to *Crazy Girls*." *New Theatre Quarterly* **21**(02): 178–94.

Massey, D. (2006). *For Space*. London, Sage.

—— (2007). *World City*. Cambridge, Polity.

Maticka-Tyndale, E., Herold, E. and Mewhinney, D. (1998). "Casual sex on spring break: intentions and behaviors of Canadian students." *Journal of Sex Research* **35**(3): 254–64.

Matthews, R. (2005). "Policing prostitution ten years on." *British Journal of Criminology* **45**(6): 877–95.

Mayhew, H. (1862). *London, Labour and the London Poor*. Bohn, Griffin.

Meah, A., Hockey, J. and Robinson, V. (2008). "What's sex got to do with it? A family-based investigation of growing up heterosexual during the twentieth century." *Sociological Review* **56**(3): 454–73.

Measham, F. C. (2004). "Play space: historical and socio-cultural reflections on drugs, licensed leisure locations, commericialisation and control." *International Journal of Drug Policy* **15**(5–6): 337–45.

Medhurst, A. (1997). "Negotiating the gnome zone: versions of suburbia in British popular culture." In R. Silverstone (ed.) *Visions of Suburbia*. London, Routledge.

Merrifield, A. (1996). "Public space: integration and exclusion in urban life." *City* **1**(5): 57–72.

Messerschmidt, J. (1987). "Feminism, criminology and the rise of the female sex 'delinquent' 1880–1930." *Crime, Law and Social Change* **11**(3): 243–63.

Mikkola, M. (2008). "Contexts and pornography." *Analysis* **68**: 316–20.

Miller, K. (2002). "Condemning the public: design and New York's new 42nd Street." *Geojournal* **58**(2/3): 139–48.

Miller, T. (2005). "A metrosexual eye on the queer guy." *GLQ: a journal of lesbian and gay studies* **11**(1): 112–17.

Miller, V. (2005) "Inter-textuality, the referential illusion, and the production of a gay ghetto." *Social & Cultural Geography* **6**(1): 61–79.

Milter, K. S. and Slade, J. (2005). "Global traffic in pornography." In L. S. Sigal (ed.) *International Exposure: perspectives on modern European pornography, 1800–2000*. New Brunswick, Rutgers University Press.

Minichiello, V., Marino, R., Khan, M. A. and Browne, K. (2003). "Alcohol and drug use in Australian male sex workers: its relationship to the safety outcome of the sex encounter." *AIDS Care-Psychological and Socio-Medical Aspects of AIDS/HIV* **15**(4): 549–61.

Mitchell, D. (1995). "The end of public space? People's park, definitions of the public, and democracy." *Annals of the Association of American Geographers* **85**(1): 108–33.

—— (2000). *Cultural Geographies – an introduction*. Oxford, Blackwells.

Moon, K. H. S. (1997). *Sex Among Allies: military prostitution in U.S.–Korea relations*. New York, Columbia University Press.

Moran, L. (2009). "Researching the irrelevant and the invisible: sexual diversity in the judiciary." *Feminist Theory* **10**(3): 281.

Morgensen, S. (2009). "Back and forth to the land: negotiating rural and urban sexuality among the radical faeries." In E. Lewin and W. Leap (eds) *Out in Public: reinventing lesbian/gay anthropology in a globalizing world*. Oxford, Blackwells.

Morris, C. and Sloop, J. (2006). "'What lips these lips have kissed': refiguring the politics of queer public kissing." *Communication and Critical/Cultural Studies* **3**(1): 1–26.
Morris, N. (2009). "Naked in nature: naturism, nature and the senses in early 20th century Britain." *Cultural Geographies* **16**(3): 283.
Morrison, C. A. (2011). "Heterosexuality and home: intimacies of space and spaces of touch, emotion." *Emotion, Space, and Society* **4**. Forthcoming.
Mort, F. (2000). *Dangerous Sexualities*. London, Routledge & Kegan Paul.
—— (2007). "Striptease: the erotic female body and live sexual entertainment in mid-twentieth-century London." *Social History* **32**(1): 27–53.
—— (2010). *Capital Affairs: London and the making of the permissive society*. New York, Yale University Press.
Mort, F. and Nead, L. (1999). "Introduction – sexual geographies." *Sexualities* **37** (Spring): 5–10.
Mulford, C., Wilson, R. E. and Parmley, A. (2009). "Geographic aspects of sex offender residency restrictions: policy and research." *Criminal Justice Policy Review* **20**(1): 3–12.
Mumford, L. (1961). *The City in History*. New York, Harcourt and Brace.
Munro, V. (2008). "Constructing consent: legislating freedom and legitimating constraint in the expression of sexual autonomy." *Akron Law Review* **41**: 923.
Munt, S. (1995). "The lesbian flaneur". In D. Bell and G. Valentine (eds) *Mapping Desire: geographies of sexualities*. London, Routledge.
Murphy, K. P. and Pierce, J. L. (eds) (2010). *Queer Twin Cities*. Minneapolis, University of Minnesota Press.
Namaste, K. (1996). "Genderbashing: sexuality, gender, and the regulation of public space." *Environment and Planning D: society and space* **14**: 221–40.
Namaste, V. (2000). *Invisible Lives: the erasure of transsexual and transgendered people*. Chicago, University of Chicago Press.
Nasaw, D. (1992). "Cities of light, landscapes of pleasure." In O. Zunz and D. Ward (eds) *The Landscape of Modernity: essays on New York City, 1900–1940*. New York, Russell Sage Foundation.
Nash, C. (2006). "Toronto's gay village (1969–82): plotting the politics of gay identity." *The Canadian Geographer/Le Géographe Canadien* **50**(1): 1–16.
—— (2010). "Trans geographies, embodiment and experience." *Gender, Place & Culture* **17**(5): 579–95.
Nash, C. and Bain, A. (2007). "Reclaiming raunch? Spatializing queer identities at Toronto women's bathhouse events." *Social & Cultural Geography* **8**(1): 47–62.
Nast, H. (1998). "Unsexy geographies." *Gender, Place & Culture* **5**(2): 191–206.
Nast, H. J. (2002). "Queer patriarchies, queer racisms, international." *Antipode* **34**(5): 874–909.
Nast, H. J. and Wilson, M. O. (1994). "Lawful transgressions: this is the house that Jackie built . . . ." *Assemblage* **24**: 48–56.
Nead, L. (1988). *Myths of Sexuality: representations of women in Victorian Britain*. Oxford, Oxford University Press.

—— (1992). *The Female Nude: art, obsecenity, and sexuality*. London, Routledge.

—— (1997). "Mapping the self: gender, space, and modernity in mid-Victorian London." *Environment and Planning A* **29**(4): 659–72.

—— (2005). *Victorian Babylon: people, streets and images in nineteenth-century London*. New Haven, Yale University Press.

Nelen, H. and Huisman, W. (2008). "Breaking the power of organized crime? The administrative approach in Amsterdam." In J. M. Nelen, D. Siegel and H. Nelen (eds) *Organized Crime: culture, markets and policies*. New York, Springer.

New York Times (1998). "Hall of shame." 23 February 1998: 5.

Nigianni, B. (2006). "'An avenue that looks like me': re-presenting the modern cityscape." *European Studies: a journal of European culture, history and politics* **23**(1): 63–80.

Nochlin, L. (2006). "Afterword." In A. M. D'Souza and T. McDonagh (eds) *The Invisible Flaneuse: gender, public space and visual culture in nineteenth century Paris*. Manchester, Manchester University Press.

Noys, B. (2008). "The end of the monarchy of sex': sexuality and contemporary nihilism." *Theory, Culture & Society* **25**(5): 104.

O'Connell Davidson, J. (2001). "The sex tourist, the expatriate, his ex-wife and her 'Other': the politics of loss, difference and desire." *Sexualities* **4**(1): 5–24.

O'Rourke, M. (2005). "On the eve of a straight queer future: notes towards an antinormative heteroerotic." *Feminism and Psychology* **15**(1): 111–16.

Obrador-Pons, P. (2007). "A haptic geography of the beach: naked bodies, vision and touch." *Social & Cultural Geography* **8**(1): 123–41.

Ogborn, M. (1993). "Law and discipline in nineteenth century English state formation: the Contagious Diseases Acts of 1864, 1866 and 1869." *Journal of Historical Sociology* **6**(1): 28–55.

Oppermann, M. (1999). "Sex tourism." *Annals of Tourism Research* **26**(2): 251–66.

Osgerby, B. (2005). "The bachelor pad as cultural icon: masculinity, consumption and interior design in American men's magazines, 1930–65." *Journal of Design History* **18**(1): 99–108.

Oswin, N. (2006). "Decentring queer globalization: diffusion and the 'global gay'." *Environment and Planning D: society and space* **24**(5): 277–90.

—— (2010). "The modern model family at home in Singapore: a queer geography." *Transactions of the Institute of British Geographers* **35**(2): 256–68.

Oswin, N. and Olund, E. (2010). "Governing intimacy." *Environment and Planning D: society and space* **28**(1): 60–67.

Overell, R. (2009). "The Pink Palace, policy and power: home-making practices and gentrification in Northcote." *Continuum* **23**(5): 681–95.

Ozbay, C. (2010). "Nocturnal queers: rent boys' masculinity in Istanbul." *Sexualities* **13**(5): 645–63.

Paasonen, S. (2009). "Healthy sex and pop porn: pornography, feminism and the Finnish context." *Sexualities* **12**(5): 586–604.

—— (2010). "Labors of love: netporn, web 2.0, and the meanings of amateurism." *New Media & Society*. **12**(8) 1297–312.

Pain, R. (1991). "Space, sexual violence and social control: integrating geographical and feminist analyses of women's fear of crime." *Progress in Human Geography* **15**(4): 415.

Papadopoulos, L. (2010). *Sexualisation of Young People: a review*. London, The Home Office.

Papayanis, M. A. (2000). "Sex and the revanchist city: zoning out pornography in New York." *Environment and Planning D: society and space* **18**(3): 341–53.

Parent-Duchâtelet, A. (1836). *De la Prostitution dans la Ville de Paris, Considérée sous le Rapport de l'Hygiène Publique, de la Morale et de l'Administration*. Paris, J.-B. Baillière.

Park, R. (1915). "The city: suggestions for the investigation of human behaviour in the city environment." *American Journal of Sociology* **20**: 577–612.

Paur, J. (2002). "Circuits of queer mobility: tourism, travel and globalization." *GLQ: a journal of lesbian and gay studies* **8**(1–2): 101–37.

Peiss, K. (1987). *Cheap Amusements: working women and leisure in turn-of-the-century New York*. Philadelphia, Temple University Press.

Peniston, W. A. (2001). "Pederasts, prostitutes, and pickpockets in Paris of the 1870s." *Journal of Homosexuality* **41**(3–4): 169–87.

Penttinen, E. (2004). *Corporeal Globalization. Narratives of Subjectivity and Otherness in the Sexscapes of Globalization*. Tapri Occasional Paper 92. Tampere.

Perreau, B. (2008). "Introduction: in/discipliner la sexualité." *EchoGeo* **5** (July/August): http://echogeo.revues.org/5923 (accessed 6 May 2011).

Phan, L., Fefferman, N., Hui, D. and Brugge, D. (2010). "Impact of street crime on Boston Chinatown." *Local Environment* **15**(5): 481–91.

Phelan, S. (2001). *Sexual Strangers: gay and lesbian dilemmas of citizenship*. Philadelphia, Temple University Press.

Phillip, J. and Dann, G. (1998). "Bar girls in central Bangkok". In M. Opperman (ed.) *Sex Tourism and Prostitution: aspects of leisure, recreation and work*. New York, Cognizant Communication.

Phillips, R. (2005). "Heterogeneous imperialism and the regulation of sexuality in British West Africa." *Journal of the History of Sexuality* **14**(3): 291–315.

—— (2006). "Unsexy geographies: heterosexuality respectability and the traveller's aid society." *ACME: an international e-journal for critical geographies* **5**: 163–88.

Phillips, R., Watt, D. and Shuttleton, D. (eds) (2000). *De-centring Sexualities: politics and representations beyond the metropolis*. London, Routledge.

Philo, C. (2005). "Sex, life, death, geography: fragmentary remarks inspired by 'Foucault's population geographies'." *Population, Space and Place* **11**(3): 325–33.

Pile, S. (1996). *The Body and the City: psychoanalysis, space, and subjectivity*. London, Routledge.

Pitman, B. (2002). "Re-mediating the spaces of reality television: America's Most Wanted and the case of Vancouver's missing women." *Environment and Planning A* **34**(1): 167–84.

Podmore, J. (2006). "Gone 'underground'? Lesbian visibility and the consolidation of queer space in Montreal." *Social & Cultural Geography* **7**(4): 595–625.

Pope, C. (2005). "The political economy of desire: geographies of female sex work in Havana, Cuba." *Journal of International Women's Studies* **6**(2): 99–118.

Potter, R. and Potter, L. (2001). "The internet, cyberporn, and sexual exploitation of children: Media moral panics and urban myths for middle-class parents?" *Sexuality & Culture* **5**(3): 31–48.

Prior, J. (2008). "Planning for sex in the city: urban governance, planning and the placement of sex industry premises in inner Sydney." *Australian Geographer* **39**(3): 339–52.

Pritchard, A. and Morgan, M. (2006). "Hotel Babylon? Exploring hotels as liminal sites of transition and transgression." *Tourism Management* **27**(5): 762–72.

Probyn, E. (1995). "Viewpoint: lesbians in space. Gender, sex and the structure of missing." *Gender, Place & Culture* **2**(1): 77–84.

Pryce, A. (2001). "'Some people live out their own snuff movie': knowledge, safer sex and managing desire in the city." *Sexual and Relationship Therapy* **16**(1): 15–34.

—— (2003). "Governmentality, the iconography of sexual disease and 'duties' of the STI clinic." *Nursing Inquiry* **8**(3): 151–61.

Puar, J. K. (2002). "Circuits of queer mobility: tourism, travel, and globalization." *GLQ: a journal of lesbian and gay studies* **8**(1): 101–37.

Quilley, S. (1997). "Constructing Manchester's 'New Urban Village': gay space in the entrepreneurial city". In G. B. Ingram, A. Bouthillette and Y. Retter (eds) *Queers in Space: communities/public places/sites of resistance*. Seattle, Bay Press.

Quinn, B. (2002). "Sexual harassment and masculinity: the power and meaning of 'girl watching'." *Gender & Society* **16**(3): 386–98.

Ravenscroft, N. and Gilchrist, P. (2009). "Spaces of transgression: governance, discipline and reworking the carnivalesque." *Leisure Studies* **28**(1): 35–49.

Reay, B. (2010). *New York Hustlers*. Manchester, Manchester University Press.

Reckless, W. (1933). *Vice in Chicago*. Chicago, University of Chicago Press.

Redoutey, E. (2009). "Ville et sexualités publiques: un essai d'ethno(géo)graphie." Unpublished PhD thesis, Paris, L'Institut d'Urbanisme de Paris.

Rendell, J. (1997). "The pursuit of pleasure: London rambling." *Culture, Theory and Critique* **40**(1): 30–41.

—— (1998). "West End rambling: gender and architectural space in London 1800–1830." *Leisure Studies* **17**(2): 108–22.

—— (2002). *The Pursuit of Pleasure*. London, Continuum.

Rendell, J., Penner, B. and Borden, I. (2000). *Gender Space Architecture: an interdisciplinary introduction*. Oxford, Spon Press.

Rich, A. (1980). "Compulsory heterosexuality and lesbian existence." *Signs* **5**(4): 631–60.

Richardson, D. (2000). "Constructing sexual citizenship: theorising sexual rights." *Critical Social Policy* **20**(1): 101–35.

Riechl, A. (2002). "Fear and lusting in Las Vegas and New York: sex, political economy, and public space". In J. Eade and C. Mele (eds) *Understanding the City: contemporary and future perspectives*. New York, Routledge.

Roberts, M. (1991). *Living in a Man-Made World: gender assumptions in modern housing design*. London, Taylor & Francis.

Roberts, M. and Eldridge, A. (2007). "Quieter, safer, cheaper? Planning for more inclusive town centres at night." *Planning Practice and Research* **22**: 253–66.

Roseneil, S. (2006). "On not living with a partner: unpicking coupledom and cohabitation." *Sociological Research Online* **11**(3): www.socresonline.org.uk/11/3/roseneil.html (accessed 6 May 2011).

Rosewarne, L. (2007). "Pin-ups in public space: sexist outdoor advertising as sexual harassment." *Women's Studies International* **30**(4): 313–25.

Ross, B. L. (2010). "Sex and (evacuation from) the city: the moral and legal regulation of sex workers in Vancouver's West End, 1975–85." *Sexualities* **13**(2): 197–218.

Ross, B. and Greenwell, K. (2005). "Spectacular striptease: performing the sexual and racial other in Vancouver, B.C., 1945–1975." *Journal of Women's History* **17**(1): 137–71.

Ross, K. (1996). *Fast Cars, Clean Bodies: decolonization and the reordering of French culture*. Cambridge, MA, MIT Press.

Rothenberg, T. (1995). "And she told two friends: lesbians creating urban social space." In D. Bell and G. Valentine (eds) *Mapping Desire: geographies of sexualities*. London, Routledge.

Rowe, D. (2003). *Representing Berlin: sexuality and the city in Imperial and Weimar Germany*. Chichester, Ashgate.

Rubin, G. (1984). "Thinking sex: notes for a radical theory of the politics of sexuality". In C. S. Vance (ed.) *Pleasure and Danger: exploring female sexuality*. London, Pandora.

Rubin, R. (2001). "Alternative lifestyles revisited, or whatever happened to swingers, group marriages, and communes?" *Journal of Family Issues* **22**(6): 711–26.

Ruffolo, D. (2009). *Post-Queer Politics*. Chichester, Ashgate.

Runstedler, C. (2006). "Magdalenes". In M. Ditmore (ed.) *Encyclopedia of Prostitution and Sex Work*. Westport, Greenwoods.

Rush, E. and La Nauze, A. (2006). *Corporate Paedophilia: sexualisation of children in Australia*. Canberra, The Australia Institute.

Rutherford, S. (1999). *Organizational Culture, Patriarchal Closure and Women Managers*. Bristol, Policy Press.

Ruting, B. (2007). *Is the Golden Mile Tarnishing? Urban and social change on Oxford Street, Sydney*. Queer Space: centres and peripheries, UTS Sydney. Unpublished conference proceedings.

Ryder, A. (2004). "The changing nature of adult entertainment districts: between a rock and a hard place or going from strength to strength?" *Urban Studies* **41**(9): 1659–86.

—— (2009). "Red light districts." In R. Hutchinson (ed.) *Encyclopedia of Urban Studies*. Forest Hills, Sage.

Sagar, T. (2005). "Street watch: concept and practice: civilian participation in street prostitution control." *British Journal of Criminology* **45**(1): 98–112.

Sanchez, L. E. (2004). "The global e-rotic subject, the ban, and the prostitute-free zone: sex work and the theory of differential exclusion." *Environment and Planning D: society and space* **22**(6): 861–83.

Sanders, T. (2005). *Sex Work: a risky business*. Cullompton, Willan.

—— (2008a). "Male sexual scripts: intimacy, sexuality and pleasure in the purchase of commercial sex." *Sociology* **42**(3): 400–17.

—— (2008b). "Selling sex in the shadow economy." *International Journal of Social Economics* **35**(10): 704–16.

—— (2009a). "Controlling the 'anti sexual' city: sexual citizenship and the disciplining of female street sex workers." *Criminology and Criminal Justice* **9**(4): 507–25.

—— (2009b). "Kerbcrawler rehabilitation programmes: curing the 'deviant' male and reinforcing the 'respectable' moral order." *Critical Social Policy* **29**(1): 77–99.

Sanders, T. and Campbell, R. (2007). "Designing out vulnerability, building in respect: violence, safety and sex work policy." *British Journal of Sociology* **58**(1): 1–19.

Sanders, T., O'Neill, M. and Pitcher, J. (2009). *Prostitution: sex work, policy and politics*. London, Sage.

Sanselme, F. (2004). "Des riverains à l'épreuve de la prostitution: fondements pratiques et symboliques de la morale publique." *Les Annales de la Recherche Urbaine* **95**: 111–17.

Sassen, S. (2002). "Global cities and survival circuits." In B. Ehrenreich and A. R. Hochschild (eds) *Global Women: nannies, maids and sex workers*. New York, Metropolitan.

Saunders, K. (1995) "Controlling (hetero)sexuality: the implementation and operation of contagious diseases legislation in Australia 1868–1945." In D. Kirkby (ed.) *Sex, Power and Justice: historical perspectives of law in Australia*. Melbourne: Oxford University Press.

Scherrer, K. (2008). "Coming to an asexual identity." *Sexualities* **11**(5): 621–41.

Schivelbusch, W. (1988). *Disenchanted Night: the industrialization of light in the nineteenth century* (trans. A. Davies). Oxford, Berg.

Schlichter, A. (2004). "Queer at last? Straight intellectuals and the desire for transgression." *GLQ: a journal of lesbian and gay studies* **10**(4): 543.

Schlör, J. (1998). *Nights in the Big City: Paris, Berlin, London 1840–1930*. London, Reaktion Books.

Sedgwick, E. (1985). *Between Men: English literature and male homosocial desire*. New York, Columbia University Press.

—— (1990). *The Epistemology of the Closet*. Berkeley, University of California Press.

—— (1993). *Tendencies*. Durham, Duke University Press.

Self, H. (2003). *The Fallen Daughters of Eve*. London, Frank Cass.

Self, R. O. (2008). "Sex in the city: the politics of sexual liberalism in Los Angeles, 1963–79." *Gender & History* **20**: 288–311.

Sennett, R. (1970). *The Uses of Disorder: personal identity and city life*. London, Knopf.

Shah, S. P. (2006). "Producing the spectacle of Kamathipura: the politics of red light visibility in Mumbai." *Cultural Dynamics* **18**(3): 269–92.

Shahani, P. (2008). *Gay Bombay: globalization, love and (be)longing in contemporary India*. Delhi: Sage Publications.

Shaver, F. (1985). "Prostitution: a critical analysis of three policy approaches." *Canadian Public Policy/Analyse de Politiques* **11**(3): 493–503.

Sheehan, J. (2004). *Skin City: behind the scenes of the Las Vegas sex industry*. London, Harper Collins.

Shields, R. (1991). *Places on the Margin*. London, Routledge.

Short, J. (2004). *Global Metropolitan: globalizing cities in a capitalist world*. New York, Routledge.

Shumsky, N. and Springer, L. (1981). "San Francisco's zone of prostitution, 1880–1934." *Journal of Historical Geography* **7**(1): 71–89.

Sibalis, M. (2004). "Urban space and homosexuality: the example of the Marais, Paris' 'gay ghetto'." *Urban Studies* **41**(9): 1739–58.

Sibley, D. (1995). *Geographies of Exclusion: society and difference in the West*. London, Routledge.

Sides, J. (2006). "Excavating the postwar sex district in San Francisco." *Journal of Urban History* **32**(3): 355–79.

—— (2009). *Erotic City: sexual revolutions and the making of modern San Francisco*. Oxford, Oxford University Press.

Sigel, L. (2002). *Governing Pleasures: pornography and social change in England, 1815–1914*. New York, Rutgers University Press.

Sigusch, V. (1998). "The neosexual revolution." *Archives of Sexual Behavior* **27**(4): 331–59.

Simmel, G. (1903). *The Metropolis and Mental Life*. Dresden, Petterman.

Simon, B. (2002). "New York Avenue: the life and death of gay spaces in Atlantic City, New Jersey 1920–90." *Journal of Urban History* **28**: 300–27.

Simpson, N. (2004). "Coming attractions: a comparative history of the Hollywood Studio System and the porn business." *Historical Journal of Film, Radio and Television* **24**(4): 635–52.

Skeggs, B. (1999). "Matter out of place: visibility and sexualities in leisure spaces." *Leisure Studies* **18**(3): 213–32.

—— (2005). "Formation: the making of class and gender through visualizing moral subject." *Sociology* **39**(5): 965–82.

Slater, D. (2004). "Trading sexpics on IRC: embodiment and authenticity on the Internet." *Body and Society* **4**(4): 91–117.

Slavin, S. (2004). "Drugs, space and sociality in a gay nightclub in Sydney." *Journal of Contemporary Ethnography* **33**(3): 265–95.

Slocum, R. (2009). "Commentary on 'public health as urban politics, urban geography: venereal biopower in Seattle, 1943–83'." *Urban Geography* **30**(1): 30–5.

Smith, A. M. (2001). "Missing poststructuralism, missing Foucault: Butler and Fraser on capitalism and the regulation of sexuality." *Social Text* **19**(2): 103–25.

Smith, B. (2010). "Scared by, of, in, and for Dubai." *Social & Cultural Geography* **11**(3): 263–83.

Smith, C. (2007a). "Designed for pleasure: style, indulgence and accessorized sex." *European Journal of Cultural Studies* **10**(2): 167–84.

—— (2007b). *One for the girls! The pleasures and practices of reading women's porn*. Bristol, Intellect Books.

—— (2010). "Pornographication: a discourse for all seasons." *International Journal of Media and Cultural Politics* **6**(1): 103–8.

Smith, D. (1999). "The civilizing process and the history of sexuality: comparing Norbert Elias and Michel Foucault." *Theory and Society* **28**(1): 79–100.

Smith, D. and Holt, L. (2005). "'Lesbian migrants in the gentrified valley' and 'other' geographies of rural gentrification." *Journal of Rural Studies* **21**(3): 313–22.

Smith, G. K. and King, M. (2009). "Naturism and sexuality: broadening our approach to sexual well-being." *Health and Place* **15**: 439–46.

Smith, J. (2004). "The politics of sexual knowledge: the origins of Ireland's containment culture and the Carrigan Report (1931)." *Journal of the History of Sexuality* **13**(2): 208–33.

Smith, J. S. (2010). "Just how naughty was Berlin? The geography of prostitution and female sexuality in Curt Moreck's *Erotic Travel Guide*." *Amsterdamer Beitrage zur neueren Germanistik* **75**(1): 53–77.

Smith, M. and Davidson, J. (2008). "Etiquette and civility". In P. Hubbard, T. Hall and J. Short (eds) *Compendium of Urban Studies*. London, Sage.

Smith, M., Grove, C. and Smith, E. (2008). "Agency-based male sex work: a descriptive focus on physical, personal, and social space." *The Journal of Men's Studies* **16**(2): 193–210.

Smith, N. (1996). *The New Urban Frontier: gentrification and the revanchist city*. New York, Routledge.

—— (2002). "New globalism, new urbanism: gentrification as global urban strategy." *Antipode* **34**(3): 427–50.

Socia, K. and Stamatel, J. P. (2010). "Assumptions and evidence behind sex offender laws: registration, community notification, and residence restrictions." *Sociology Compass* **4**(1): 1–20.

Solnit, R. (2001). *Wanderlust: a history of walking*. New York, Penguin Group US.

Solomon, S. E., Rothblum, E. D. and Balsam, K. F. (2005). "Money, housework, sex, and conflict: same-sex couples in civil unions, those not in civil unions, and heterosexual married siblings." *Sex Roles* **52**(9): 561–75.

Sonmez, S., Apostolopoulos, Y., Yu, C. H., Yang, S., Mattila, A. and Yu, L. (2006). "Binge drinking and casual sex on spring break." *Annals of Tourism Research* **33**(4): 895–917.

Spencer, D. (2009). "Sex offender as homo sacer." *Punishment & Society* **11**(2): 219–40.

Stakelbeck, F. and Frank, U. (2003). "From perversion to sexual identity: concepts of homosexuality and its treatment in Germany." *Journal of Gay & Lesbian Psychotherapy* **7**(1): 23–46.

Stella, F. (2008). "Homophobia begins at home: lesbian and bisexual women's experiences of the parental household in urban Russia." *Kul'tura* **2**(1): 12–17.

—— (2011). "The politics of in/visibility: lesbian sexuality, urban space and collective agency in Ulíianovsk, Russian Federation." *Europe-Asia Studies* **63**(3). Forthcoming.

Storr, M. (2002). "Classy lingerie." *Feminist Review* **71**: 18–36.

Stryker, S. (2008a). "Dungeon intimacies: the poetics of transsexual sadomasochism." *Parallax* **14**(1): 36–47.

—— (2008b). "Transgender history, homonormativity, and disciplinarity." *Radical History Review* **100**: 145.

Stryker, S. and Whittle, S. (2006). *The Transgender Reader*. New York: Routledge Press.

Sullivan, B. (2010). "When (some) prostitution is legal: the impact of law reform on sex work in Australia." *Journal of Law and Society* **37**(1): 85–104.

Sutton, B. (2007). "Naked protest: memories of bodies and resistance at the World Social Forum." *Journal of International Women's Studies* **8**: 139.

Svanström, Y. (2000). *Policing Public Women. The regulation of prostitution in Stockholm 1812–1880*. Stockholm: Atlas Akademi.

Swanson, G. (2007). *Drunk with the Glitter: space, consumption and sexual instability in modern urban culture*. London, Routledge.

Symanski, R. (1981). *The Immoral Landscape*. Toronto, Butterworths.

Takahashi, L. (1997). "The socio-spatial stigmatization of homelessness and HIV/AIDS: toward an explanation of the NIMBY syndrome." *Social Science & Medicine* **45**(6): 903–14.

Takeyama, A. (2005). "Commodified romance in a Tokyo host club." In M. McLelland and R. Gasgupta (eds) *Genders, Transgenders and Sexualities in Japan*. London, Routledge Curzon.

Tambe, A. (2009). *Codes of Misconduct: regulating prostitution in late colonial Bombay*. Minneapolis, University of Minnesota Press.

Tatar, M. (1995). *Lustmord: sexual murder in Weimar Germany*. Princeton, Princeton University Press.

Taulke-Johnson, R. (2010). "Assertion, regulation and consent: gay students, straight flatmates, and the (hetero) sexualisation of university accommodation space." *Gender and Education* **9**(1): 1–17.

Tewksbury, R. (1990). "Patrons of porn: research notes on the clientele of adult bookstores." *Deviant Behavior* **11**(3): 259–71.

—— (1996). "Cruising for sex in public places: the structure and language of men's hidden, erotic worlds." *Deviant Behavior* **17**(1): 1–19.

—— (2010). "Men and erotic oases." *Sociology Compass* **12**(4): 1011–19.

Thomas, M. (2004). "Pleasure and propriety: teen girls and the practice of straight space." *Environment and Planning D: society and space* **22**(5): 773–90.

Thompson, E. and Morgan, E. (2008). "'Mostly straight' young women: variations in sexual behavior and identity development." *Developmental Psychology* **44**(1): 15–21.

Thomson, S. A. (2007). "It's moments like these you need mint: a mapping of spatialised sexuality in Brisbane." *Queensland Review* **14**(2): 93–104.

Thurnell-Read, T. P. (2009). *Masculinity, Tourism and Transgression: a qualitative study of British stag tourism in an Eastern European City*. Unpublished PhD thesis, Department of Sociology, University of Warwick.

Tibbals, C. A. (2011). "Sex work, office work: women working behind the scenes in the US Adult Film Audience." *Gender, Work and Organisation*. Forthcoming.

Tomsen, S. and Markwell, K. (2009). "Violence, cultural display and the suspension of sexual prejudice." *Sexuality & Culture* **13**(4): 201–17.

Tomso, G. (2008). "Viral sex and the politics of life." *South Atlantic Quarterly* **107**(2): 265.

Tonnies, F. (1887). *Community and Society*. Leipzig, Fues's Verlag.

Tucker, A. (2009). *Queer Visibilities: space, identity and interaction in Cape Town*. London, Routledge.

Tucker, D. M. (2007). "Preventing the secondary effects of adult entertainment establishments: is zoning the solution?" *Journal of Land Use and Environmental Law* **12**(2): 385–430.

Turner, M. (2003). *Backward Glances: cruising the queer streets of New York and London*. London, Reaktion.

Tye, D. and Powers, A. (1998). "Gender, resistance and play: bachelorette parties in Atlantic Canada." *Women's Studies International* **21**(5): 551–61.

Uebel, M. (2004). "Striptopia?" *Social Semiotics* **14**: 3–19.

Valentine, G. (1993). "(Hetero)sexing space: lesbian perceptions and experiences of everyday spaces." *Environment and Planning D: society and space* **11**(3): 395–411.

—— (1996). "Lesbian productions of space." In N. Duncan (ed.) *BodySpace: destabilizing geographies of gender and sexuality*. London, Routledge.

—— (1997). "Lesbian separatist communities in the United States." In P. Cloke and J. Little (eds) *Contested Countryside Cultures: otherness, marginalisation, and rurality*. London, Routledge.

—— (2006). "Globalizing intimacy: the role of information and communication technologies in maintaining and creating relationships." *Women's Studies Quarterly* **34**: 365–93.

—— (2008). "The ties that bind: towards geographies of intimacy." *Geography Compass* **2**(6): 2097–110.

Valentine, G. and Skelton, T. (2003). "Finding oneself, losing oneself: the lesbian and gay 'scene' as a paradoxical space." *International Journal of Urban and Regional Research* **27**(4): 849–66.

Valverde, M. (1999). "The harms of sex and the risks of breasts: obscenity and indecency in Canadian law." *Social Legal Studies* **8**(2): 181–97.

Valverde, M. and Cirak, M. (2003). "Governing bodies, creating gay spaces. policing and security issues in 'gay' downtown Toronto." *British Journal of Criminology* **43**(1): 102.

Van der Veen, M. (2000). "Beyond slavery and capitalism: producing class difference in the sex industry". In J. K. Gibson-Graham, S. Resnick and R. D. Wolff (eds) *Class and its Others*. Minneapolis, University of Minnesota Press.

Van Doorninck, M. and Campbell, R. (2006). "Zoning street sex work: the way forward." In R. Campbell and M. O'Neill (eds) *Sex Work Now*. Cullompton, Willan.

Van Every, J. (1996). "Sinking into his arms . . . arms in his sink: heterosexuality and feminism revisited." In L. Adkins and V. Merchant (eds) *Sexualising the Social*. London, Macmillan.

Van Heyningen, E. B. (1984). "The social evil in the Cape colony 1868–1902: prostitution and the contagious diseases acts." *Journal of Southern African Studies* **10**(2):170–97.

Vasudevan, A. (2006). "Experimental urbanisms: psychotechnik in Weimar Berlin." *Environment and Planning D: society and space* **24**(6): 799.

Verbraeck, H. (1990). "The German Bridge: a street hookers' strip in the Amsterdam Red Light District." In Lambert, Y. (ed.) *The Collection and Interpretation of Data from Hidden Populations*. Rockville, US Dept of Health.

Visser, G. (2003). "Gay men, leisure space and South African cities: the case of Cape Town." *Geoforum* **34**(1): 123–37.

—— (2008). "The homonormalisation of white heterosexual leisure spaces in Bloemfontein, South Africa." *Geoforum* **39**: 1344–58.

Visser, J. (1998). "Selling private sex in public spaces: street sex work in the Netherlands." Paper presented at Perspectives on female prostitution conference, Liverpool Hope University.

Voss, G. (2007). "The dynamics of technological change in a stigmatised sector." *Science and Technology Policy Research Unit*. Falmer, University of Sussex.

Wagenaar, H. (2006). "Democracy and prostitution: deliberating the legalization of brothels in the Netherlands." *Adminstration and Society* **38**: 198–225.

Wagenaar, H. and Altink, S. (2009). "To toe the line: streetwalking as contested space." In D. Canter (ed.) *Safer Sex in the City: the experience and management of street prostitution*. Chichester, Ashgate.

Walby, K. (2009). "Ottawa's National Capital Commission Conservation Officers and the policing of public park sex." *Surveillance & Society* **6**(4): 367.

Walkowitz, J. (1982). *Prostitution and Victorian Society*. Cambridge, Cambridge University Press.

—— (1998). "Going public: shopping, street harassment, and streetwalking in late Victorian London." *Representations* **62**: 1–30.

Wallace, A. (2007). "The geography of girl watching in postwar Montreal." *Space and Culture* **10**(3): 349.

Walsh, K. (2007). "'It got very debauched, very Dubai!' Heterosexual intimacy amongst single British expatriates." *Social & Cultural Geography* **8**: 507–33.

Wan, L. (2009). "Citizen media action and the transformation of indecency and obscenity censorship in Hong Kong." *International Journal of Media and Cultural Politics* **5**(1&2): 131–7.

Ward, H., Day, S. and Webber, J. (1999). "Risky business: health and safety in the sex industry over a 9 year period." *Sexually Transmitted Infections* **75**(5): 340.

Ward, J. (2001). *Weimar Surfaces: urban visual culture in 1920s Germany*. Los Angeles, University of California Press.

Warner, M. (1991). "Fear of a queer planet." *Social Text* **29**: 3–17.

—— (2000). *The Trouble with Normal*. New York, Free Press.

Warren, C. (1974). *Identity and Community in the Gay World*. New York, Johns Hopkins Press.

Watson, S. (2006). *City Publics: the (dis) enchantments of urban encounters*. London, Routledge.

Watt, P. (1998). "Going out of town: youth, race, and place in the South East of England." *Environment and Planning D: society and space* **16**: 687–704.

Weeks, J. (1985). *Sexuality and its Discontents: meanings, myths, and modern sexualities*. London, Routledge and Kegan Paul.

—— (2007). *The World we have Won: the remaking of erotic and intimate life*. London, Routledge.

Weeks, J., Heaphy, B. and Donovan, C. (2001). *Same Sex Intimacies: families of choice and other life experiments*. New York, Routledge.

Weightman, B. (1980). "Gay bars as private places." *Landscape* **24**(1): 9–17.

Weiss, M. (2008). "Gay shame and BDSM pride: neoliberalism, privacy, and sexual politics." *Radical History Review* **100**: 87.

Weitman, S. (1999). "On the elementary forms of the socioerotic life." In M. Featherstone (ed.) *Love and Eroticism*. London, Sage.

Weitzer, R. (2010). "Sex work: paradigms and policies." In R. Weitzer (ed.) *Sex for Sale: prostitution, pornography and the sex industry* (second edition). New York, Routledge.

West, M. D. (2002). *Japanese Love Hotels: legal change, social change and industry change*. Chicago, John M. Olin Center for Law & Economics, University of Michigan.

Westhaver, R. (2005). "Coming out of your skin: circuit parties, pleasure and the subject." *Sexualities* **8**(3) 347–74.

Weston, K. (1995). "Get thee to a big city: sexual imaginary and the great gay migration." *GLQ: a journal of lesbian and gay studies* **2**(3): 253–77.

White, C. (2006). "The Spanner trials and the changing law on sadomasochism in the UK." *Journal of Homosexuality* **50**(2): 167–87.

—— (2007). "'Save us from the womanly man': the transformation of the body on the beach in Sydney, 1810 to 1910." *Men and Masculinities* **10**(1): 22.

White, M. (2010). "What a mess: eBay's narratives about personalization, heterosexuality, and disordered homes." *Journal of Consumer Culture* **10**(1): 80–104.

Whitzman, C. (2007). "Stuck at the front door: gender, fear of crime and the challenge of creating safer space." *Environment and Planning A* **39**(11): 2715–32.

Whowell, M. (2010). "Male sex work: exploring regulation in England and Wales." *Journal of Law & Society* **37**(1): 125–44.

Wilkinson, E. (2009). "Perverting visual pleasure: representing sadomasochism." *Sexualities* **12**(2): 181–98.

—— (2010). "What's queer about non-monogamy now?" In M. Barker and D. Langdridge (eds) *Understanding Non-monogamies*. London, Routledge.

Williams, L. (1989). *Hard Core: power, pleasure and the 'frenzy of the visible'*. Los Angeles, University of California Press.

Williams, S., Lyons, L. and Ford, M. (2008). "It's about bang for your buck, bro: Singaporean men's online conversations about sex in Batam, Indonesia." *Asian Studies Review* **32**(1): 77–97.

Wilson, E. (1995). *The Sphinx in the City: urban life, the control of disorder, and women*. London, Virago.

Wilton, R. (1998). "The constitution of difference: space and psyche in landscapes of exclusion." *Geoforum* **29**(2): 173–85.

Winship, J. (2000). "Women outdoors: advertising, controversy and disputing feminism in the 1990s." *International Journal of Cultural Studies* **3**(1) 27–55.

Wolff, J. (1985). "The invisible flâneuse. women and the literature of modernity." *Theory, Culture & Society* **2**(3): 37–45.

Wolkowitz, J. (2006). *Bodies at Work*. London, Sage.

Wonders, N. A. and Michalowski, B. (2001). "Bodies, borders, and sex tourism in a globalized world: a tale of two cities – Amsterdam and Havana." *Social Problems* **48**(4): 545–71.

Wong, T. and Yeoh, B. (2003). "Fertility and the family: an overview of pro-natalist population policies in Singapore." *Asian MetaCentre for Population and Sustainable Development Analysis Research Paper Series* **12**: 1–25.

Worthington, B. (2005). "Sex and shunting: contrasting aspects of serious leisure within the tourism industry." *Tourist Studies* **5**(3): 225–46.

Wosick-Correa, K. R. and Joseph, L. J. (2008). "Sexy ladies sexing ladies: women as consumers in strip clubs." *Journal of Sex Research* **45**(3): 201–16.

Wouters, C. (1987). "Developments in the behavioural codes between the sexes: the formalization of informalization in the Netherlands, 1930–85." *Theory, Culture & Society* **4**(2): 405.

Wunneburger, D., Olivares, M. and Maghelal, P. (2008). "Internal security for communities: a spatial analysis of the effectiveness of sex offender laws." In D. Z. Sui (ed.) *Geospatial Technologies and Homeland Security*. New York, Springer.

Yang, Y. (2006). *Whispers and Moans: interviews with the men and women of Hong Kong's sex industry*. Hong Kong, Blacksmith Books.

Young, Iris Marion (1990). *Justice and the Politics of Difference*. Princeton, NJ: Princeton University Press.

Zandbergen, P. A. and Hart, T. C. (2009). "Restrictions for sex offenders geocoding accuracy considerations in determining residency." *Criminal Justice Policy Review* **20**(1): 62–90.

Zelizer, V. A. (2007). *The Purchase of Intimacy*. New York, Princeton University Press.

Zheng, T. (2009). *Red Lights: the lives of sex workers in postsocialist China*. Minneapolis, University of Minnesota Press.

Zook, M. (2003). "Underground globalization: mapping the space of flows of the Internet adult industry." *Environment and Planning A* **35**: 1261–86.

—— (2007). "Report on the location of the Internet adult industry." In K. Jacobs, M. Janssen and M. Pasquinelli (eds) *C'lick Me: a netporn studies reader*. Amsterdam, Institute of Network Culture.

Zukin, S. (1995). "Urban lifestyles: diversity and standardisation in spaces of consumption." *Urban Studies* **35**(5–6): 825.

Index